LUBRICATED WEAR
Science and Technology

Vol. 6	Friction and Wear of Polymers (Bartenev and Lavrentev)
Vol. 10	Microstructure and Wear of Materials (Zum Gahr)
Vol. 11	Fluid Film Lubrication – Osborne Reynolds Centenary (Dowson et al., Editors)
Vol. 12	Interface Dynamics (Dowson et al., Editors)
Vol. 13	Tribology of Miniature Systems (Rymuza)
Vol. 14	Tribological Design of Machine Elements (Dowson et al., Editors)
Vol. 15	Encyclopedia of Tribology (Kajdas et al.)
Vol. 16	Tribology of Plastic Materials (Yamaguchi)
Vol. 17	Mechanics of Coatings (Dowson et al., Editors)
Vol. 18	Vehicle Tribology (Dowson et al., Editors)
Vol. 19	Rheology and Elastohydrodynamic Lubrication (Jacobson)
Vol. 20	Materials for Tribology (Glaeser)
Vol. 21	Wear Particles: From the Cradle to the Grave (Dowson et al., Editors)
Vol. 22	Hydrostatic Lubrication (Bassani and Piccigallo)
Vol. 23	Lubricants and Special Fluids (Stepina and Vesely)
Vol. 24	Engineering Tribology (Stachowiak and Batchelor)
Vol. 25	Thin Films in Tribology (Dowson et al., Editors)
Vol. 26	Engine Tribology (Taylor, Editor)
Vol. 27	Dissipative Processes in Tribology (Dowson et al., Editors)
Vol. 28	Coatings Tribology – Properties, Techniques and Applications in Surface Engineering (Holmberg and Matthews)
Vol. 29	Friction Surface Phenomena (Shpenkov)
Vol. 30	Lubricants and Lubrication (Dowson et al., Editors)
Vol. 31	The Third Body Concept: Interpretation of Tribological Phenomena (Dowson et al., Editors)
Vol. 32	Elastohydrodynamics – '96: Fundamentals and Applications in Lubrication and Traction (Dowson et al., Editors)
Vol. 33	Hydrodynamic Lubrication – Bearings and Thrust Bearings (Frêne et al.)
Vol. 34	Tribology for Energy Conservation (Dowson et al., Editors)
Vol. 35	Molybdenum Disulphide Lubrication (Lansdown)
Vol. 36	Lubrication at the Frontier – The Role of the Interface and Surface Layers in the Thin Film and Boundary Regime (Dowson et al., Editors)
Vol. 37	Multilevel Methods in Lubrication (Venner and Lubrecht)
Vol. 38	Thinning Films and Tribological Interfaces (Dowson et al., Editors)
Vol. 39	Tribological Research: From Model Experiment to Industrial Problem (Dalmaz et al., Editors)
Vol. 40	Boundary and Mixed Lubrication : Science and Applications (Dowson et al., Editors)
Vol. 41	Tribological Research and Design for Engineering Systems (Dowson et al., Editors)

TRIBOLOGY SERIES, 42
EDITOR: D. DOWSON

LUBRICATED WEAR
Science and Technology

A. SETHURAMIAH
ITMMEC, Indian Institute of Technology
New Delhi, India

2003

ELSEVIER
Amsterdam – Boston – London – New York – Oxford – Paris
San Diego – San Francisco – Singapore – Sydney – Tokyo

ELSEVIER SCIENCE B.V.
Sara Burgerhartstraat 25
P.O. Box 211, 1000 AE Amsterdam, The Netherlands

First edition 2003

Library of Congress Cataloging in Publication Data
A catalog record from the Library of Congress has been applied for.

British Library Cataloguing in Publication Data
A catalogue record from the British Library has been applied for.

ISBN: 0-444-51092-3 (vol. 42)
ISSN: 0-444-41677-3 (series)

♾ The paper used in this publication meets the requirements of ANSI/NISO Z39.48-1992 (Permanence of Paper).
Printed in Hungary.

Preface

Wear in lubricated contacts and the associated problems of running-in and scuffing are of increasing importance in modern machinery. This is a consequence of the modern compact designs that increase the operating severity. Though lubricated contacts are more important industrially, emphasis so far has been on dry wear and several books have been written on this topic. The information related to lubricated wear is scattered and is mainly available in journals. The reported investigations in turn seem to emphasise one aspect or the other of the wear process. The author felt the need to treat lubricated wear in a consolidated manner. The present book is the outcome of the effort in this direction.

The book is aimed at all those interested in lubricated wear including the lubricant formulators, user industries, and basic researchers in tribology. Besides consolidation of the available information the major purpose of the book is to strengthen the theory-practice interface. Such a focus identifies the gaps in knowledge that need to be filled and will be of interest to tribologists. Development of knowledge is a long-term goal and the industry will be interested in the more immediate improvements to practice. This is addressed in the book and well-argued methodologies are suggested that can ameliorate the present situation.

The developments in practice will depend on the effective dialogue between engineers and scientists from various disciplines. This dialogue is only possible when all aspects related to lubricated wear are well appreciated. The first chapter gives a perspective of tribology to those who have major interest in the lubricant chemistry with secondary interest in tribology. The second chapter provides an overview of lubricant technology and will be useful to tribologists who need an appreciation of the complexity of a formulated lubricant, and the limits within which it has to function.

The subsequent chapters deal with wear and the associated problems of scuffing, running-in and fatigue. The starting point is the third chapter devoted to dry wear. The fourth and fifth chapters deal with mechanisms of boundary lubrication and wear modelling of metallic materials. The sixth chapter considers the dry and lubricated wear of polymers and ceramics, which is of growing importance. The seventh chapter deals with laboratory evaluations. Fatigue and wear in mixed lubrication is covered in the eighth chapter. The final chapter deals with the important issue of relevance of laboratory tests to real situations.

The coverage in each chapter has to be necessarily limited and it was a difficult task to decide on the extent of coverage. The main approach adopted is to clarify the central concepts involved and then critically examine their utility. The necessary background needed with regard to temperature rise, contact mechanics, adsorption, and chemical reactions is provided where appropriate. The coverage in these areas is not extensive but adequate to serve the main objective of the book. Detailed theoretical coverage of hydrodynamic and mixed lubrication is not within the scope of this book. A reasonable attempt has however been made to explain the central ideas and how boundary contact is affected in mixed lubrication. Adequate references are provided so that the interested reader can pursue any topic in more detail. References and nomenclature have been given chapter-wise.

Some of the new aspects considered in the book may interest the reader. These are given below.

- The role of chemical reactions in wear and scuffing is examined from a different perspective. This leads to a new proposal for antiwear mechanism of chemical additives.
- The low wear rates encountered in lubricated conditions can only be judged by effective quantification of running-in and steady state wear. A methodology based on experimental research is presented and its utility discussed.
- A new approach to use polymer composites for hydrogenerator thrust pads is presented based on experimental research. This extends the potential use of polymeric materials for heavy-duty applications under lubricated conditions.
- Asperity level conformity can sometimes occur in the contact. Experimental research conducted on this aspect brings out the importance of such conformity in reducing metal contact in mixed lubrication.
- Boundary lubrication has an important role to play in metal forming operations and is difficult to simulate. A new experimental approach based on oblique plastic impact is presented.
- Measurement of low levels of wear in industrial components is of importance. An experimental method based on bearing length curves is presented to measure liner wear.

- The wear map approach has been emphasised and its utility in improving wear evaluation has been given a detailed consideration.

In the final analysis a book of this type has to judged on the basis of its impact on practice. Any improvements to practice in the near future will be a source of satisfaction to the author.

Acknowledgments

I wish to express my thanks to all those who have contributed to the successful completion of the book. First and foremost I wish to thank Professor Duncan Dowson who in the capacity of editor of the Tribology series, made valuable comments with regard to all the chapters. His contribution has been of significant importance in shaping the book. I am also thankful to Mr. Dean Eastbury who has promptly responded to all my questions regarding the publication aspect.

Several examples of the work elaborated in the book are based on my research and industrial experience spread over 30 years. Many colleagues and students have contributed to this effort. During my stay at IIT, Delhi I had the good fortune of interacting with Professor U. R. K. Rao, Dr. C. R. Jagga, Dr. Braham Prakash, and Dr. Om Prakash who collaborated with me in academic and industrial research. Their interaction helped me to sharpen my focus on the theory-practice interface for which I am thankful to them. I am also thankful to Professor B. C. Nakra who inspired me to join IIT, Delhi. The stay at IIT has helped me to develop ideas at a deeper level.

My thanks are also due to Dr. Rajesh Kumar who has recently completed his PhD and actively participated in all aspects of the book writing. His research contributions on running-in have an important place in the book. Many of my former students have contributed to lubricated wear with their academic work. Their contributions have an important place in the present volume. The interaction with these students, and shaping of ideas with them has been the most satisfying part of my career. I take this opportunity to thank Doctors Mange Ram Tyagi, V. R. K. Sastry, Rathin Kumar Banerjee,. S. K. Karmakar, R. Prakash, and T. R. Choudhary for their direct and indirect contribution to this volume.

I am also thankful to several other tribologists who have interacted with me at conferences and meetings. These interactions have been useful in exchanging ideas. The interesting discussions I had with Professor K. C. Ludema on several occasions have been of particular relevance to this book.

Finalisation of the book and printing are time consuming tasks. In this connection I wish to express my sincere thanks to Mr. M. Satyanarayana, CFO, of Nipuna Services Ltd for providing me with state-of-the-art facilities to complete the job quickly.

Turning to the more distant past, my own doctoral research was conducted at the Tokyo Institute of Technology during 1969-73 under the able guidance of late Professor T. Sakurai. He has been instrumental in shaping me into a full-fledged researcher. My first exposure to lubrication research was at the French Petroleum Institute where I was a trainee during 1962-63. This training played an important role in my later development.

I also like to express my thanks to my son, Ravindra who helped me with the Autocad drawings. My daughter, Sushma helped me with issues related to permissions and the latest technical information. Last but not the least I wish to express my warm thanks to my wife, Jayashree who did her best to let me work without hindrance. Without her continued support this book would not have been possible.

Contents

1. Tribology in perspective

1.1 Introduction

This chapter is intended for those readers who have expertise in lubrication science and technology with secondary interest in tribology. Though many specialised books are available in the area, it is considered appropriate to introduce the main ideas in a consolidated manner. This will equip the reader for the specialised chapters on boundary lubrication and wear.

Tribology may be defined as the science and technology of surfaces in relative motion. It encompasses the commonly known aspects of friction, wear and lubrication. Such a definition emphasises the fact that friction, wear, and lubrication are interconnected and need to be studied together. Tribology can also be considered as an enabling technology as technological development depends on the satisfactory resolution of tribological problems. In the present chapter, the laws of friction are first stated and the conceptual explanation of these laws is given. The limitations of the simple model are considered followed by model that takes into account the growth in contact area. Surface roughness and contact models are considered in the next section and their implications to tribology are discussed. The issue of surface temperature rise is considered next. This is followed by a consideration of friction of non-metallic materials. The final section deals with lubrication. Boundary lubrication and wear are only mentioned where appropriate as separate chapters are devoted to these subjects.

1.2 Laws of friction and their explanation

The origin of friction has interested several investigators since 14th century. Historical development of the ideas involved is well documented by Dowson [1]. The major earlier investigators included Amontons, Coulomb, and Leonardo de Vinci. The laws of friction are generally associated with the name of Amontons, which state that the friction force in a sliding contact is proportional to the normal load and is independent of the geometric area of contact. Friction force is the tangential force necessary to overcome friction. Experimental studies by several researchers have shown that these laws are approximately valid. Many attempts have been made in the past to explain these laws without much success. The first satisfactory explanation was due to Bowden and Tabor [2]. Their model provides a conceptual explanation of the laws of friction. They considered that when two rough surfaces are brought into contact under normal load the real contact is made at some asperity peaks only. The pressure at these contacts is very high and plastic flow at these asperities is assumed. The flow pressure is taken as the hardness of the metal. They further assumed that the shear strength of the real junctions is equal to the shear strength of the material. The contact model for two identical metals is given in Fig. 1.1. On this basis the laws of friction can be explained as follows:

$$A_r = \frac{L}{H} \tag{1.1}$$

$$F = s_m A_r \tag{1.2}$$

$$f = \frac{F}{L} = \frac{s_m}{H} \tag{1.3}$$

where

A_r	= real contact area, m^2
L	= load, N
H	= hardness, N/m^2
F	= friction force, N
s_m	= shear strength of metal, N/m^2
f	= friction coefficient

From Eq. (1.3) it can be seen that f is a constant since s_m and H are material properties. It follows that friction force is proportional to load. As real area is governed only by hardness and load, it is independent of geometric area.

The above explanation is based on contact of metallic materials. When the two contacting surfaces are not identical s_m and H are taken to be those of the softer material. Since s_m is nearly equal to 1/6 of the hardness for a metallic material it follows that f is about 0.16 and is a constant irrespective of the material combination. Both these observations are not validated in practice. Surfaces in normal atmosphere can have f values ranging from 0.3 to 1.2 and depend on the material combination. One aspect of practical interest is that friction coefficient for a given material pair is reasonably constant over a range of operating conditions.

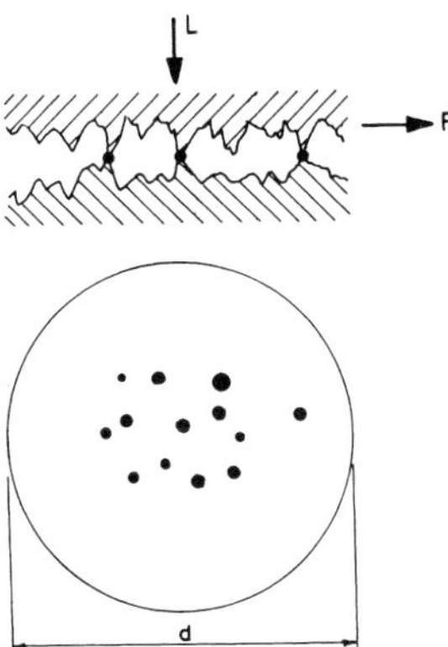

Fig. 1.1. Real area spots within circular geometric area of diameter d.

The value of f however depends on the material pair, surface condition, and environment.

The problem was recognised by Tabor [3] and he provided a plausible explanation for the phenomenon. The earlier model proposed considered yielding is only due to normal force. In reality, yielding occurs under combined normal and tangential stresses. Assuming plane strain conditions, the relation proposed for plastic deformation is

$$p^2 + \alpha s_i^{\,2} = \alpha s_m^{\,2} \tag{1.4}$$

where p is the normal pressure, α is a constant, s_i is the shear strength of the interface, and s_m is the shear strength of the metal.

From this equation, two important observations can be made. For a given load L, the normal pressure p decreases and tends to zero as s_i approaches s_m. This happens because the real area increases and hence the normal pressure decreases. If $\frac{s_i}{s_m} = \kappa$ and $\alpha = 9$ the following equation can be derived for friction coefficient f as given below:

$$f = \sqrt{\frac{\kappa^2}{9(1-\kappa^2)}} \tag{1.5}$$

The above equation shows that for very clean surfaces with κ approaching 1.0, f can reach very high values. However, even for a minor contamination with say, $\kappa = 0.95$ f reaches a modest value of 1.0. Experimental verification of these ideas is quite extensive. With very clean surfaces generated under high vacuum in the range of 10^{-6} to 10^{-8} torr f values greater than 10 can be obtained for several metals. It is usual to refer to the asperity contacts as adhesive junctions and the theory involved as adhesion theory of friction.

The above theory cannot be quantified as κ cannot be defined for a tribological contact. While high friction can be reconciled with junction growth, plastic deformation of the junctions additionally contributes to friction. Hence, the values differ from metal to metal depending on issues like ductility, work hardening, and atomic arrangement on the surface. The atomic arrangement is of importance since shear strength depends on the crystallographic plane on which sliding occurs and can be well demonstrated for single crystals [4]. Thus prediction of f from the above equation is not possible even for a given κ. Yet another problem with the theory is that metals with hexagonal close packed (hcp) structure do not show strong junction growth and slide easily even with very clean surfaces as shown for cobalt alloys [5]. Despite these limitations, the qualitative aspect of the theory is of importance. In space applications tribological contacts operate under very high vacuum. Under such condition surfaces will no longer have protective oxides and are prone to high friction and seizure. These contacts can be operated only with thin lead or solid lubricant films interposed between the surfaces that eliminate direct asperity contact.

The normal engineering surfaces are generally covered with oxides that reduce the direct metallic contact. This is an advantage from a tribological point of view. The protection from oxide films depends on several factors like composition, adherence to surface, and the balance between removal and formation rates. The asperity contacts in real systems can range from elastic to plastic contacts as will be seen in the later section. Even when the contacts are predominantly elastic and covered with protective oxides the f values can range between 0.3 and 1.2 for most of the metals.

1.2.1 Other aspects of friction

Besides adhesion and plastic deformation, it is recognised that ploughing which results in plastic grooving can also contribute to friction. Ploughing can occur when hard asperities of one surface groove the softer surface or when abrasive particles come between the surfaces. The ploughing action due to asperities depends on their shape and hardness. Models are available to estimate the deformation losses when a cone or a sphere sinks into a surface and displaces the material ahead by plastic deformation while sliding [6,7]. Models for ploughing are based on a sinking depth needed to support the load and the tangential force needed to displace the material ahead. The load support area is based on half the vertical projected area while the tangential force is based on the horizontal projected area of the groove. For engineering surfaces, the ploughing component of friction is generally small. This is because the hemispherical asperities with radii in the range of 100-1000 μm have negligible ploughing action. Ploughing is important in operations like grinding and cutting. Hard wear particles or abrasive dust particles can also contribute to ploughing.

The explanation of friction so far has not distinguished between static and dynamic friction. Static friction refers to the value at the onset of sliding while kinetic friction refers to the value during gross sliding. Kinetic friction is usually smaller than static friction. When kinetic friction is lower than the momentary static friction oscillations are possible which is referred to as stick-slip. Stick-slip depends on the elastic response of the system, and the variation of kinetic friction with velocity. Stick-slip problems can arise in components like brakes leading to vibration and noise. Careful control of materials in such cases is important to overcome the problem.

In the case of pure rolling between hard metals the energy losses involved are mainly due to elastic hysteresis. The energy losses for such cases are very small and when expressed in terms of a friction coefficient are < 0.01. In the case of elastic materials like rubber tires, the energy losses are substantial and f values for automobile tires range from 0.2 to 0.8.

1.3 Contact of surfaces

On a microscopic scale, the surfaces are not atomically smooth and have undulations. These undulations are referred to as roughness. The roughness can be traced by a fine diamond stylus that follows the surface profile. A typical roughness trace is illustrated in Fig. 1.2. In this profile, the high points, called asperities, appear sharp because the vertical magnification is much higher than the horizontal magnification. When the scaling in both directions is the same, asperities will have gentle slopes and this should be kept in mind. Surfaces can have waviness due to manufacturing processes and this waviness is filtered by cut-off filter. This ensures that the parameters calculated from the profile are representative of the roughness. Several roughness parameters are then obtained by processing the digitised data with microprocessor. Centre line average (R_a) and rms value (R_q) are based on a mean line that divides the roughness profiles into equal areas. The measured heights at different points of the profile, below and above the mean line, are processed to obtain these values in the given sampling length. In common engineering practice R_a is usually specified for the surfaces while the rms value is common for statistical analysis of the surfaces. R_q is usually referred to as σ in contact analysis. The R_a values depend on the nature of application and typically range from 0.1 to 2.0μm. The other parameters that can be routinely obtained by the instrument include asperities per unit length, average asperity slope and curvature. The values as obtained are two dimensional along a chosen line.

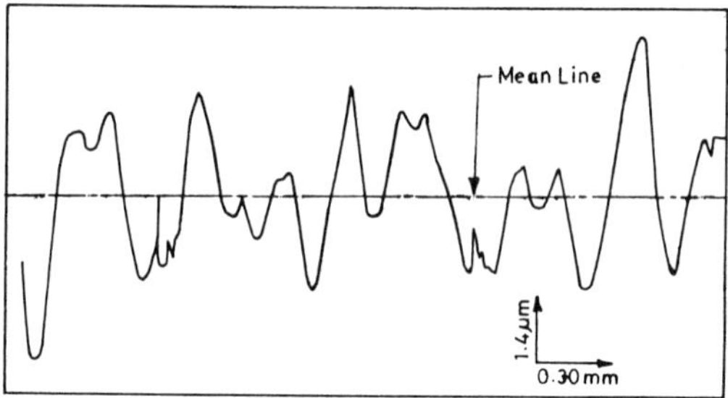

Fig. 1.2. Roughness profile of cast iron surface with R_a= 0.88 μm.

The three dimensional parameters are necessary to characterise the contact. For isotropic surfaces, designating the high spots as summits and, those in the profile trace as peaks the following assumptions may be made for a surface with Gaussian height distribution [8]:

1. The average curvature of summits is nearly same as that obtained for peaks.

2. The density of summits (asperities) per unit area η is approximately equal to $1.8\,\eta_p^2$ where η_p is the number of peaks per unit length.
3. The distribution of summits is Gaussian with rms value same as that for general height distribution.

Two rough surfaces, which are nominally flat, may now be brought into contact. It is easier to visualise the problem in terms of a rough surface contacting a rigid flat surface. As asperity contacts depend on the nature of height distribution, it is necessary to specify the distribution function. The present treatment is based on a Gaussian distribution that reasonably approximates many engineering surfaces. The concept is not restricted and can be applied to other kinds of distribution as well provided the distribution functions are known. Fig. 1.3a shows the nature of general height distribution and the cumulative height distribution. Fig.1.3b shows the contact situation. The shaded area indicates the contact zone. When a rough surface is brought into contact with a smooth rigid flat under load highest asperities first come in contact. An asperity with height z_i will be compressed by $(z_i - d)$ where d is the distance from mean line and the rigid flat. As the load increases asperities with lower heights come into contact to support the load. The problem is to find the real contact area as a function of load assuming elastic contact. Following Greenwood and Williamson [9] the following integrals can be obtained:

$$n = \eta A_n \int_d^\infty \varphi(z)dz \tag{1.6}$$

$$A_r = \pi\beta\eta A_n \int_d^\infty (z-d)\varphi(z)dz \tag{1.7}$$

$$L = \tfrac{4}{3} E\ \beta^{0.5}\eta A_n \int_d^\infty (z-d)^{3/2}\varphi(z)dz \tag{1.8}$$

where

n	= number of contacts
A_r	= total real area of contacts, m^2
L	= total load, N
η	= asperity density = number of asperities per m^2
A_n	= nominal contact area, m^2
β	= mean asperity radius, m
d	= separation between rigid plane and mean surface height, m
z	= asperity height over mean surface height line, m

$\varphi(z)$ = Gaussian height distribution function = $\dfrac{1}{\sigma\sqrt{2\pi}}\exp\left(-\dfrac{z^2}{2\sigma^2}\right)$

E $=\left(\dfrac{1-\nu_1^{\ 2}}{E_1}+\dfrac{1-\nu_2^{\ 2}}{E_2}\right)^{-1}$, N/m^2 where $\nu_{1,2}$ and $E_{1,2}$ refer to Poisson ratio and elasticity modulus of the two surfaces.

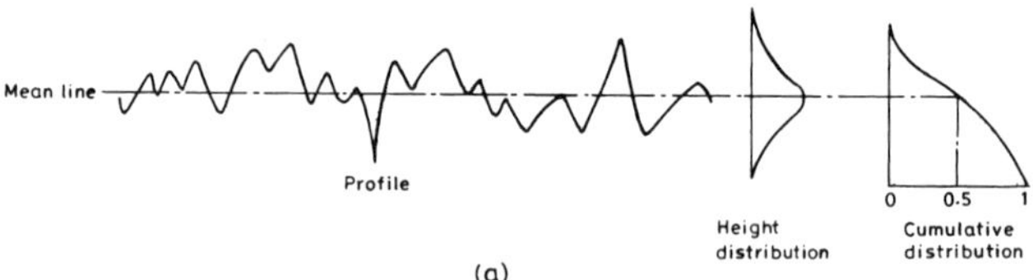

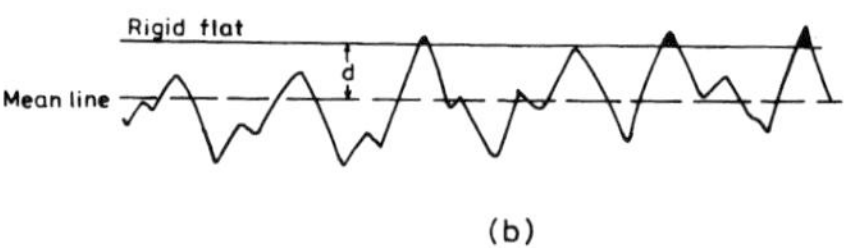

Fig. 1.3. (a) Roughness trace and Gaussian and cumulative height distribution; (b) Loaded contact between rigid flat and rough surface with separation d.

For the case of rigid flat contacting a rough surface E will be governed by Poisson ratio and elastic modulus of the rough surface only. The Eq. (1.6) expresses the number of asperities in contact on the basis that all those asperities exceeding d will be in contact. The expressions for real area and load are obtained from basic contact mechanics relationships for spherical contact. They relate the deformation depth to load and contact area for a known β. For a given asperity 'i' the Eqs. (1.9a) and (1.9b) give these relationships as follows:

$$A_i = \pi\delta_i\beta \tag{1.9a}$$

$$L_i = \tfrac{4}{3}E\ \beta^{1/2}\delta_i^{\ 3/2} \tag{1.9b}$$

where A_i is the area , L_i is the load on asperity with height z_i, and δ_i is the deformation equal to $(z_i - d)$.

$\varphi(z)$ is known for a Gaussian distribution and the above equations can be numerically integrated. The major conclusion from this modelling is that the real area is nearly proportional to the load even if the contacts are elastic. Also real area is nearly independent of geometric area.

The actual contact is between two rough surfaces. The usual approach is to treat the contact as an equivalent surface against a rigid flat. The effective modulus E is as defined under Eq. (1.8). Equivalent radius β^*, and equivalent roughness σ^* are defined in Eqs. (1.10a) and (1.10b) as

$$\beta^* = \frac{\beta_1 \beta_2}{\beta_1 + \beta_2} \tag{1.10a}$$

$$\sigma^* = \sqrt{\sigma_1^2 + \sigma_2^2} \tag{1.10b}$$

where β_1 and β_2 are the asperity radii of the two surfaces and σ_1 and σ_2 are the rms roughness of the two surfaces. In the case of above contact model with rigid smooth surface β^* and σ^* will be governed by β and σ of the rough surface only.

1.3.1 Implications of the contact model

The laws of friction derived for plastic contact postulate that real area is proportional to load and is independent of the geometric area. This is also valid for elastic contacts. Hence, it is considered that laws of friction are also applicable to elastic contacts. It is however difficult to define the adhesive forces for elastic contacts. Their values are expected to be lower than in the case of plastically deforming asperities. On the other hand, for a given load, the real area is higher compared to the plastic case and the overall *f* value may not be very different. The argument provided here is only directional. The growth in contact area is difficult to envisage for elastic contacts and the real area may be governed only by the normal load. The issue of adhesion is important with regard to wear also and is elaborated in chapter 3.

The deformation of an asperity contact will become plastic beyond a critical deformation depth. On this basis, plasticity index has been proposed in [9] to characterise the deformation as

$$\psi = \frac{E}{H}\sqrt{\frac{\sigma^*}{\beta^*}} \tag{1.11}$$

where H is the hardness of the softer material. It has been proposed that for $\psi <$ 0.6 the overall contact is essentially elastic while for values of $\psi > 1.0$ the contact is predominantly plastic.

As an example let two identical mild steel surfaces with $E_{1,2}$=203 GPa, H=1.8 GPa, $\sigma_{1,2}$=0.2 μm, and $\beta_{1,2}$=500 μm be brought into contact. Based on earlier equations for equivalent values ψ is calculated as 2.08. This value indicates the surfaces predominantly yield plastically. Similar surfaces made of hard steel with H=7.0GPa will have a ψ value of 0.57 and the contact will be essentially elastic. Thus, plasticity index can be used for an approximate assessment of the contact situation.

The above model is for static contact. During sliding, the load progressively shifts from one set of asperities to another. It is considered, statistically speaking, that the contact situation repeats itself. This may not be true particularly for cases with significant plastic deformation. Strong junctions with contact growth may upset the load sharing and some of them may contribute to friction due to residual adhesion even when the load is released in the contact. It may also be observed that the model depends on average parameters. Some individual contacts may have values far different from average and these may be critical in the initial adjustment of surfaces normally referred to as running-in. It is also known that the roughness parameters as obtained are a function of sampling interval, the cut-off selected, and the stylus dimensions. Current research based on detailed formulation of random surfaces and fractals [10,11] may eventually provide better contact models. Another issue is the scale of observation. For example, if the surface roughness is obtained by atomic force microscopy many smooth areas observed with stylus meter will show undulations. The fundamental question then is what is roughness. From a practical point of view, one has to be content with the explanation of a phenomenon based on observation at a relevant scale. Awareness of this problem is essential since explanations of mechanisms in tribology are based on observations at different scales.

The contact model has important implications to wear. The removal process is governed by the nature of contact and is far more sensitive to the surface condition than friction. The f value for metals ranges from 0.3 to 1.2 in normal atmosphere as stated earlier. On the other hand, the wear rates can vary by several orders of magnitude. Boundary lubrication can have a dramatic influence on wear. A drop of mineral oil introduced between two rubbing surfaces can bring down friction by a factor of 10 while the wear may reduce by a factor of 1000 or more.

1.4 Surface temperatures in sliding contact

The frictional heating in tribological contacts has a significant influence on the performance. The heat generated due to friction is conducted away into the solids and there is a rise in temperature at the interface over and above the bulk temperature of the solids. This temperature can affect the surface oxidation, lubricant failure, chemical reaction with additives, and a host of related problems. The temperature rise is calculated theoretically as it is very difficult to measure these temperatures.

1.4.1 Theoretical calculation

Jaegar [12] presented an analysis of moving and stationary heat sources that forms the basis for most of the work in this area. When steady state is assumed, which is normally the case, simple final equations are available to analyse the temperature problem. In the present analysis, these equations are used and they provide adequate background to appreciate the problem. When steady state conditions do not occur the original equations of Jaegar are to be applied.

The temperature rise can be analysed based on a square projection of one body sliding on the other as shown in Fig. 1.4. This projection is making contact over the whole geometric contact area of $4l^2$. Both the bodies are considered semi-infinite for this analysis. Body II is moving over body I with a velocity v. Body I, which is stationary receives heat from a *moving* heat source while body I that is moving, receives heat from a *stationary* heat source. It is now required to estimate the surface temperature rise in the contact. The following nomenclature is adopted:

c_p = specific heat (J/Kg°C)
k = thermal conductivity (W/m°C)
l = half contact dimension of heat source (m)
q = heat flux (W/m^2)

v	= sliding velocity (m/s)
$\Delta\theta$	= average temperature rise (°C)
χ	= thermal diffusivity (m^2/s)
P_e	$= \frac{vl}{2\chi}$, Peclet number, non-dimensional
L	= load (Kg)

Two equations form the basis for calculating average temperature rise in contact and are based on simplification of the original theory. For a stationary heat source average temperature rise is given by

$$\Delta\theta = 0.946\left(\frac{ql}{k}\right) \tag{1.12}$$

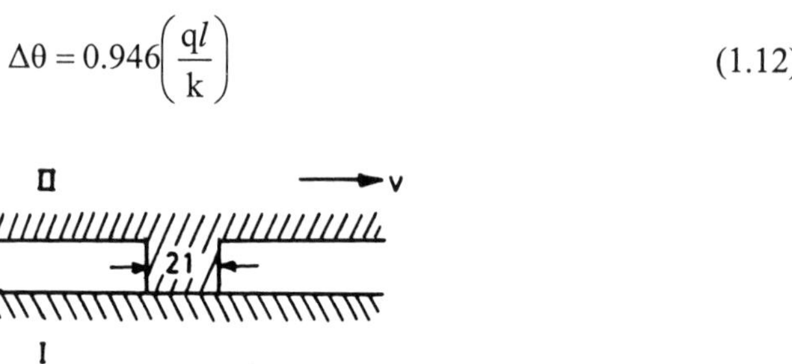

Fig. 1.4. Square projection of body II moving over stationary surface I with velocity v.

This equation is also applicable for slow speed contact with $P_e < 0.5$. For a moving heat source with $P_e > 5.0$ the mean temperature rise is

$$\Delta\theta = 0.75\left(\frac{ql}{k}\right)P_e^{-1/2} \tag{1.13}$$

The case for Peclet number ranging from 0.5 to 5.0 shall be treated later.

The distribution of temperature in the square contact is not uniform and varies from $-l$ to $+l$. The position of the maximum temperature varies with P_e. The simplified equations refer to an integrated average value over the contact. For stationary contact, the maximum temperature is $\cong$ 1.15 times the average value while for moving source with $P_e > 5.0$ the maximum value is nearly 1.5 times the average value. The basic equations above refer to heat flow into one body or the other. In sliding contact, the heat generated is distributed into both the bodies II and I. The common interface is considered to have the same temperature. This requirement means that the heat flux q generated shall be so partitioned that $\Delta\theta$ is the same for both the bodies. The heat flowing into the two bodies is q_1 and q_2 respectively. Define a partition coefficient φ which is the fraction of total heat q entering the body I. Then $(1-\varphi)$ is the fraction of heat going into body II. On this basis, surface temperature rise for slow and fast moving contacts can be estimated as given below.

For slow moving contacts with $P_e < 0.5$, stationary contact theory is applicable to both the bodies. Since $\Delta\theta$ is the same for both the bodies it follows

$$\Delta\theta = 0.946\left(\frac{q\varphi l}{k_I}\right) = 0.946\left(\frac{q(1-\varphi)l}{k_{II}}\right) \tag{1.14}$$

where k_I and k_{II} refer to the thermal conductivities of the two bodies I and II. Taking $k^* = \dfrac{k_I}{k_{II}}$ it can be shown

$$\Delta\theta = 0.946\left(\frac{ql}{k_I}\right)\left(1+k^*\right)^{-1} \tag{1.15}$$

Similarly average temperature rise for cases with $P_e > 5.0$ is

$$\Delta\theta = \frac{0.75\left(\dfrac{ql}{k_I}\right)P_e^{-1/2}}{1+0.793k^*P_e^{-1/2}} \tag{1.16}$$

Another approach frequently used to estimate temperature rise is to assume the *total* heat flux q enters one body or the other and calculate temperature rise for body I and body II as $\Delta\theta_I$ and $\Delta\theta_{II}$ from the corresponding equations for moving and stationary sources. The average surface temperature rise $\Delta\theta$ is then given by

$$\frac{1}{\Delta\theta} = \frac{1}{\Delta\theta_I} + \frac{1}{\Delta\theta_{II}} \tag{1.17}$$

Both the approaches above are equivalent.

Theoretically for semi-infinite bodies with 0°C bulk temperature the surface temperature is equal to the temperature rise in contact. With finite sized contacts the surface temperature, also called contact temperature is

$$\text{Contact temperature } (T_c) = \text{Bulk temperature } (T_b) + \Delta\theta \qquad (1.18)$$

The above approach is applicable to lubricated contacts where bulk temperatures of both surfaces tend to be equal. Unequal bulk temperatures are a complex problem as the heat flow into the two bodies is modified which in turn affects the temperature rise [13]. The other complications that arise include variations in thermal properties due to temperature as well as influence of oxide and other films on heat transfer [14].

The above equations are for square contacts that are also used for circular contacts replacing l by the radius of contact. Also, the square geometric area is assumed to make full contact over the geometric area. In a tribological contact, the real area is much smaller as discussed earlier. The temperature at asperities needs to be estimated besides the overall temperature rise in contact. This problem is addressed in chapter 3 considering that the asperity temperature rise is superimposed on the temperature rise over the geometric area. The asperity temperatures can be significantly higher than the temperatures based on the geometric area. The duration of asperity contacts on the other hand is short and it is not easy to decide which temperature is governing the tribological mechanism involved.

The case for Peclet numbers in the range of 0.5 to 5.0 is treated by a graphical procedure [12]. The temperature rise for the stationary source as applicable to body II is used as earlier to obtain $\Delta\theta_{II}$ assuming all heat is flowing to body II. For body I receiving heat from a moving source $\Delta\theta_I$ is given by

$$\Delta\theta_I = 0.946\left(\frac{ql}{k_I}\right)\varepsilon \qquad (1.19a)$$

This equation is the same as for stationary source multiplied by a coefficient ε. ε in turn is obtained from

$$\varepsilon = 0.398 y P_e^{-1} \qquad (1.19b)$$

where y is the ordinate value from Fig. 1.5 corresponding to the calculated P_e.

Another approach suggested by Greenwood [15] involves an interpolation method according to which the temperature rise is estimated by the following equation:

$$\frac{1}{\Delta\theta^2} = \frac{1}{\Delta\theta_I^{\ 2}} + \frac{1}{\Delta\theta_{II}^{\ 2}} \tag{1.20}$$

It may be noted that $\Delta\theta_I$ in this case is directly obtained from Eq. (1.13) at the corresponding P_e. As this calculation does not involve the graphical procedure, it is more convenient.

The above procedure can be illustrated by the following example. Consider a stationary square pin of area $1.81\text{x}10^{-5}\ \text{m}^2$. This pin is in sliding contact with a disk rotating with a velocity of 5.11 m/s. The load applied is 100 N. Assume *f* is 0.4. The pin and the disc are made of the same steel with the following thermal properties:

Thermal conductivity (k) = 43 W/m°C
Thermal diffusivity (χ) = $1.172\text{x}10^{-5}\ \text{m}^2/\text{s}$
The problem is to find the surface temperature rise based on geometric contact area.

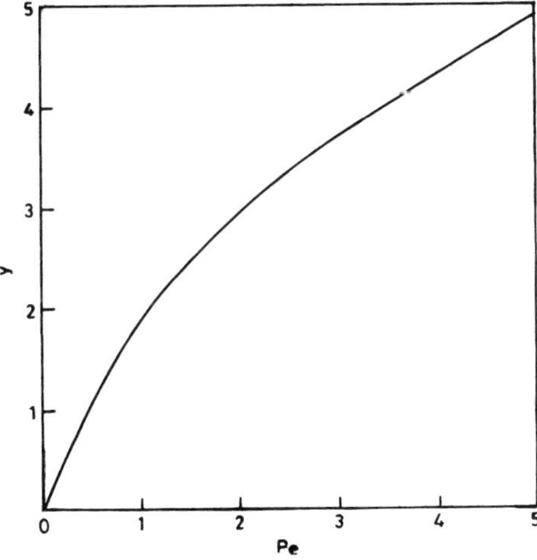

Fig. 1.5. 'y' as a function of Peclet number.

For the square contact $l = \sqrt{\dfrac{1.81x10^{-5}}{4}} = 2.127x10^{-3}\,m$

k and χ are same for both the bodies. The heat flux is taken as the frictional work done per unit time and unit area. Also since Nm/s = Watt (W)

$$q = \frac{f\,Lv}{area} = \frac{0.4x100x5.11}{1.81x10^{-5}} = 1.12x10^{7}\,W/m^{2}$$

$$P_e = \frac{vl}{2\chi} = 463.7$$

As $P_e > 5.0$, the high speed equation (1.13) is applicable. Assuming all heat is flowing to the surface I

$$\Delta\theta_I = 0.75\left(\frac{ql}{k}\right)P_e^{-1/2} = 18.71°C$$

For body II receiving heat from stationary source Eq.(1.12) is applicable. So

$$\Delta\theta_{II} = 0.946\frac{ql}{k} = 525\,°C$$

Now temperature rise $\Delta\theta$ is obtained from

$$\frac{1}{\Delta\theta} = \frac{1}{18.71} + \frac{1}{525}$$

$$So,\ \Delta\theta = 18.07\,°C$$

It is of interest to note that when Peclet number is high the interface temperature rise is governed by the moving source theory.

Thermal properties affect the temperature significantly. If for example the steel pin is replaced by a glass pin with k = 1.25 W/m°C and $\chi = 0.08x10^{-5}\,m^2/s$ the temperature rise will be as high as 130°C.

1.5 Friction of non-metals

This short section deals with the issues of friction related to ceramics and polymers. Ceramics are considered first followed by polymers. This coverage also includes an overview of the different categories of the materials used.

1.5.1 Ceramics

Engineering ceramics are being increasingly used for metal cutting tools, metal forming dies, rolling element bearings, and other applications. Ceramic coatings are also used for wear resistance. Their main advantages include low density, high strength and corrosion resistance. They also retain their strength to very high temperatures unlike metals and this makes them particularly suited for high temperature applications. The major disadvantage of ceramics is their low fracture toughness. This can be improved to some extent by controlling the grain size, and incorporating suitable sintering aids. Fabrication of ceramics involves compacting and sintering processes and careful control is necessary to obtain uniform structures. Machining of ceramics also presents problems due to their high strength.

The oxide ceramics normally used include alumina ($Al_2 O_3$) and zirconia (ZrO_2) while the non-oxide ceramics include silicon carbide (SiC), silicon nitride (Si_3N_4), tungsten carbide (WC), titanium carbide (TiC), and titanium nitride (TiN). Sialons are a special class of ceramics in which nitrogen in silicon nitride is partially replaced by Al and oxygen. The basic contact mechanics developed for metals can be used to study ceramic contact as well. Two different kind of pairing can be envisaged. The first is ceramic Vs ceramic. When the contacts are elastic ceramics have f values in the range of 0.1 to 0.3 in normal atmosphere. When contact conditions are severe, leading to fracture of ceramics at micro or macro level the hard debris increases the f values to a higher range of 0.5 to 0.8. At high temperatures, the friction of ceramics is higher in comparison to room temperature as the adsorbed species are desorbed. The second pairing is ceramic Vs metal. In this case, metal can be transferred to the ceramic and the friction changes to the metal-metal pair. The metals can also undergo oxidation by interacting with oxygen. A detailed consideration of ceramic tribology is available in an edited book [16].

Friction of some ceramics is sensitive to the environment. Silicon nitride is an interesting example. This ceramic is known to react strongly with water forming oxides and hydroxides that reduce friction significantly. A recent example where friction of the nitride is studied directly in water may be cited here [17]. The friction reduces due to surface interaction and then reaches negligible levels as the highly polished surface promotes hydrodynamic effects. Friction of alumina is also sensitive to water vapour and friction reduction occurs because of hydroxide formation [18]

For many ceramic combinations, wear tends to be high in sliding. The major cause is the fracture at micro or macro levels governed by the tensile stresses at the surface. Under moderate conditions wear can be controlled by lubricants with approaches available for metals. For high temperature applications novel ideas are necessary as liquid lubricants cannot function.

1.5.2 Polymer friction

Polymers are being used for bearing applications in automobile, aircraft, and in food and textile industries. Special applications for polymers include space and artificial human joints. Many polymers can be easily shaped and this is an advantage. Polymers have the disadvantage of low strength and limited temperature capability. These limitations restrict their use to milder operations in comparison to metals.

Thermoplastics form the major class for tribological applications. The commonly used materials include high density polyethylene (HDPE), polyamide (Nylon 6-6), polyoxymethylene (Acetal), and polytetrafluoroethylene (PTFE). Frictional behaviour of these polymers has been studied extensively. The studies have been conducted with metallic counter faces as well as glass. Two important characteristics govern the polymer friction. One is the polymer transfer to the counter face while the other is the viscoelastic behaviour during sliding. Hence, unlike metals friction can vary significantly with operating conditions. With polymers that have highly linear configuration the friction is lower. This is the case with all the above materials except nylon, which has bulky side groups.

PTFE is one of the most widely used low friction materials. The friction coefficient for PTFE varies between 0.06 and 0.2 depending on the operating conditions. Higher stress in contact tends to reduce the friction coefficient of this material. PTFE friction is governed by transfer of a thin layer on the counter face that gets oriented in the direction of sliding. Once oriented the friction becomes low with PTFE sliding against itself. The transferred layers are weakly attached to the counter face and are easily worn out. Thus, wear rates of PTFE are high and are not acceptable. Wear rates can be effectively reduced by fillers like brass, copper oxide and graphite. It is common to use two or more fillers together for effective performance. Such polymer composites can reduce wear by 2 to 3 orders of magnitude. Some of the fillers also strengthen the polymer. The friction of the composite materials is only marginally higher than pure PTFE. PTFE based materials can be used up to 250°C as its melting point is 327°C.

Nylon friction is higher than PTFE and f varies in the range of 0.4 to 0.6. The transfer to the counter face is lumpy unlike the thin oriented layers of PTFE. The transferred material is more strongly attached to the counter face and wear of this polymer is lower and acceptable in many applications. Nylon can also be impregnated with fillers to control its wear further. The operating temperature limit for nylons is around 200°C.

Normal low density polyethylene has lumpy transfer and has high friction. It also has high wear in comparison to nylon. HDPE on the other hand has lower friction with f ranging from 0.2 to 0.3. Layered transfer occurs in this case like PTFE though the layers are relatively thicker than for PTFE. The wear of HDPE is significantly lower than PTFE. The low melting temperature of 120°C for polyethylenes restricts their application significantly. The most important polymer used for biological applications in this class is the ultra high molecular weight polyethylene (UHMWP) with typical molecular weight of $4x10^6$, which is about ten times the normal molecular weight of HDPE. UHMWPE is biocompatible with good strength and low wear rate. It also has low f in the range of 0.1 to 0.2. UHMWPE also has better abrasion resistance than HDPE.

Acetal also exhibits low f values in the range of 0.15 to 0.3, with wear rates comparable to those of nylon, and can be used to temperatures up to 140°C. Several other thermoplastics like polymethylmethacrylate (PMMA), and polyphenylene sulphide (PPS) are also used for tribological applications and the above discussion was limited to the more commonly used materials. Two specialised books available cover the tribology of polymers in detail [19,20]. Development of composites with different short and long fibres as well as granular materials is receiving increased attention. Besides polymers, metal matrix composites are also gaining in importance. An edited book on the tribolgy of composites is available dealing with both polymer and metal matrix composites [21].

Thermosetting polymers are also used for bearing applications with or without fillers. Thermosetting bearings are more difficult to fabricate. These are cross-linked polymers, which decompose at high temperatures unlike thermoplastics that melt at a specific temperature. The major classes used include phenolics, epoxy resins, and polyimides. These bearing materials with various fillers are used for several applications like rolling mill bearings where they are successful in replacing metallic bearings. Phenolics are also used in brake materials as binder along with several fillers. The operating limits of polymeric materials are getting higher both in terms of strength as well as temperature by developing fibre based composites and use of high temperature polymers like polyimides and polyetherether ketone (PEEK). It is likely that in future many applications involving metallic bearings may be replaced by polymer composites, which can run dry or lubricated. The bearings are of different types starting from simple bushes to porous materials with impregnated polymers. While large body of available literature can give rough criteria for selection, it is imperative that effective simulation studies be carried out for selecting the materials. For example, development of polymer pads for hydrogenerator thrust pads conducted by Choudhary et al [22] involved simulation in a thrust bearing rig to ensure performance. Control of friction in brakes, tires, clutches and other components is of practical importance and in all such cases simulation is the only way to ensure performance. The technological aspects of friction are considered in detail by Blau [23]. The major role of the available knowledge base in tribology is the insight it provides in understanding system specific behaviour.

Polymer operating limits are normally specified in terms of pv factor where p is the pressure and v is the velocity. Limits are specified for continuous operation. This value is related to the temperature limit for the material as pv is roughly related to temperature rise in the contact. As discussed in the section on surface temperature, the temperature rise is related to P_e and for high values of Peclet number, the temperature rise should be related to $v^{1/2}$. A better approach to the thermal problem has been presented by Ettles et al [24].

1.6 Lubrication

In a broader sense, lubrication may be defined as the complete or partial separation of surfaces by interposed films. The present section is confined to liquid lubricants only. Complete separation can be affected through fluid films whose thickness is higher than the asperity dimensions. Such separation can be affected by hydrodynamic or elastohydrodynamic lubrication (EHL). Pressure generation and load support in these cases is due to hydrodynamic action where viscosity plays a major role. At the other extreme load may be supported by molecular films with no hydrodynamic load support. This regime of lubrication is referred to as boundary lubrication. In boundary lubrication, lubricant-solid interactions play the major role. The lubrication can also involve both hydrodynamic and boundary effects. Such a regime is referred to as mixed lubrication regime. This section first deals with viscosity followed by hydrodynamic lubrication and EHL. The final part of the section is limited to few observations on mixed and boundary lubrication. Many specialised books are available dealing with hydrodynamic lubrication and EHL. Treatment of this vast area in a section can only provide an appreciation of the principles involved. EHL is also referred to as EHD lubrication.

1.6.1 Viscosity

Lubricants offer resistance to shear that can be characterised by viscosity. Viscous flow is analogous to movement in a pack of cards. The lowest card is stationary while the top card at height z moves with a velocity u. Viscosity η may be expressed as

$$\eta = \frac{\tau}{s} \tag{1.21}$$

where

τ = shear stress, N/m^2

s = shear strain rate du/dz, s^{-1}

In this equation η is constant provided shear stress is proportional to shear strain. Such behaviour is termed Newtonian. Most of the lubricating oils behave as Newtonian fluids at modest pressures and shear rates commonly encountered in hydrodynamic lubrication.

Viscosity is expressed in absolute units ($N.s/m^2$) in the above equation. Many viscometers measure kinematic viscosity which is defined as viscosity /force density and has units of m^2/s. When expressed as cm^2/s the kinematic viscosity is called stoke. Centistoke (cSt) is 0.01stoke and is commonly used in characterising viscosity in industry and has units of mm^2/s.

Viscosity varies significantly with temperature. The temperature dependence can be approximated by the following relation

$$\eta = Ae^{b^*/T} \tag{1.22}$$

where, A and b^* are constants for the given fluid

The important equation used in terms of cSt (Walther equation) is expressed as

$$\log\log\left(v + c'\right) = a' - b'(\log T) \tag{1.23}$$

where v is kinematic viscosity in cSt, and a', b' and c' are constants for a given fluid. When fluid viscosity is lower than 2.0 cSt the value of c' varies. Normally one deals with fluid viscosities greater than 2.0 cSt.

This equation is considered accurate and widely used industrially. This relation forms the basis for ASTM D341. ASTM refers to American Society for Testing and Materials. In this method, a graph is scaled on the X and Y axes as per the above equation. If viscosity is known at any two temperatures, a line can be drawn passing through these points and viscosity at any other temperature can be obtained by interpolating or extrapolating from this line. The slope of the line is indicative of the variation of viscosity with temperature.

1.6.1.1 Viscosity index

Viscosity index (VI) is a comparative number that characterises the rate of change of viscosity with temperature. The kinematic viscosity in cSt of the test lubricant is compared with two reference oils, which are arbitrarily designated as 0, and 100 VI fluids. For oils with VI up to and including 100, VI is defined as

$$VI = \frac{\overline{L} - U_t}{\overline{L} - \overline{H}} x100 \tag{1.24a}$$

where

U_t = viscosity of the test fluid at 40°C

$\overline{L}$ = viscosity at 40°C of the reference fluid with 0 VI and having the same viscosity as test fluid at 100°C

$\overline{H}$ = viscosity at 40°C of the reference fluid with 100 VI and having the same viscosity as test fluid at 100°C

Higher the VI lower the change in viscosity with temperature. 100 VI oils are Pennsylvania oils while 0 VI oils are Texas naphthenic oils. To calculate the VI the viscosity of the test fluid is first measured at 40°C and 100°C. Then the required 0 and 100 VI fluids that have the same viscosity as the test fluid are selected from available tables. The same tables provide also the corresponding viscosities at 40°C for the reference fluids that are the $\overline{L}$ and the $\overline{H}$ values needed. Detailed methodology is available in ASTM D2270.

Viscosity index is widely used to characterise lubricants as will be seen in chapter 2. Petroleum lubricants manufactured until eighties had VI values in the range of 90-100. Modern lubricants can have VI values in the range of 100-150. The equation used to find VI for oils with VI greater than 100 is

$$VI = \left[\left\{\left(\text{antilog}\overline{N}\right) - 1\right\} / 0.00715\right] + 100 \tag{1.24b}$$

where

$$\overline{N} = \left(\log\overline{H} - \log U_t\right) / \log Y_t$$

In the above U_t and Y_t are the kinematic viscosities of the test fluid at 40°C and 100°C respectively. $\overline{H}$ is the same as defined in Eq. (1.24a).

1.6.1.2 Pressure effect

Viscosity also changes with pressure. Barus equation is a good approximation for the pressure influence and is expressed as

$$\eta = \eta_0 e^{\xi p} \tag{1.25}$$

where

η_0 = absolute viscosity at the given temperature and $p = 0$
ξ = pressure coefficient of viscosity, m^2/N at the given temperature
η = viscosity at pressure p

More accurate expressions for temperature and pressure effects are based on Roelands equations [25]. From a practical point of view, the pressure effect on viscosity is significant only at high pressures as in the case of line and point contacts. This is because of the value of ξ which typically ranges from 1.0-2.0×10^{-8} m^2/N. Also, at high pressures, the fluid behaves in a non-Newtonian manner. Its behaviour may be approximated by a glassy solid with a limiting shear stress.

1.6.2 Hydrodynamic lubrication

The flow of a lubricant through a convergent wedge generates pressure that can support a load. The mechanism involved is hydrodynamic in nature and this type of lubrication is referred to as hydrodynamic lubrication. This lubrication is responsible for load support in thrust and journal bearings. The principle may be illustrated by a slider bearing as shown in Fig.1.6. When the lubricant flows through the convergent wedge, the mass flow rate should be the same. This is possible only through pressure generation that in turn modifies the velocity distribution across the film in such a way that the overall flow is the same. The velocity profile at the inlet and outlet are shown in Fig. 1.6. Thrust bearings which support axial load utilise this principle. The rotating runner is supported on a set of thrust pads which are tapered. The thrust pads can be fixed or pivoted. In pivoted pads the gap changes as a function of operating conditions.

In the case of journal bearing which supports the radial load the lubricant flows through a convergent-divergent wedge. During operation, the shaft is eccentric with reference to the stationary journal. A typical full journal bearing is shown in Fig. 1.7. Eccentricity e is shown in the figure. Radial clearance c is defined as the difference in the radii of the journal and the bearing.

The flow of fluid in the bearing is governed by the well-known Reynolds equation that forms the basis for design. The normally used equation for two-dimensional flow is

$$\frac{\partial}{\partial x}\left(h^3\frac{\partial p}{\partial x}\right)+\frac{\partial}{\partial y}\left(h^3\frac{\partial p}{\partial y}\right)=12\bar{u}\eta_0\frac{\partial h}{\partial x} \tag{1.26}$$

where

h = film thickness at (x, y)

$\bar{u}$ $=\frac{u_1+u_2}{2}$ where u_1 and u_2 are the velocities of the two surfaces

This equation is based on incompressible flow and the assumption that flow is in X and Y directions only. The flow in the Y direction (into the paper) is also called side leakage. In some cases, flow in the Z direction also is to be considered in which case the three-dimensional equation is necessary. Such situations arise in dynamically loaded bearings and porous bearings. If the equation is treated in one dimension with no flow in Y direction (infinite dimension in Y direction) and applied to the slider bearing in Fig. 1.6 the equation simplifies to

$$\frac{dp}{dx}=6\bar{u}\eta_0\left(\frac{h-\bar{h}}{h^3}\right) \tag{1.27}$$

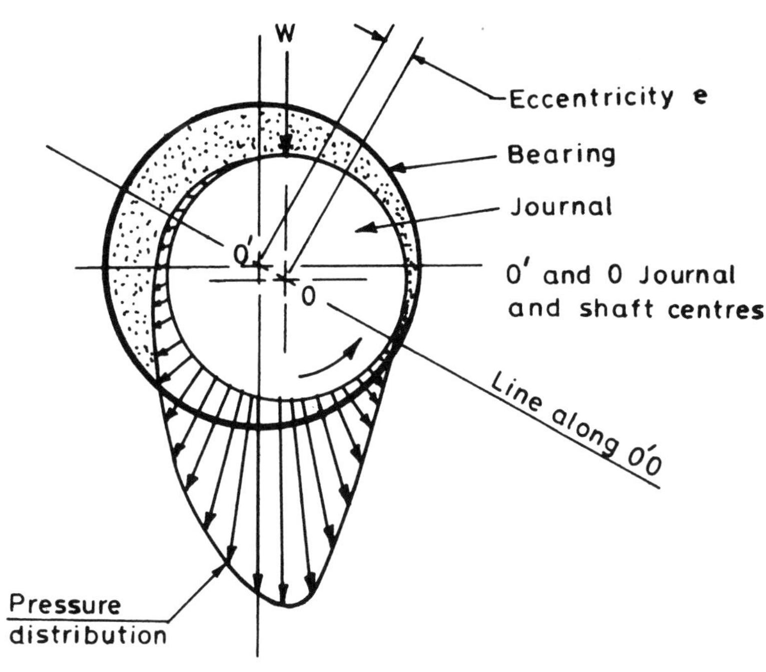

Fig.1.6. Slider bearing showing velocity and pressure distribution.

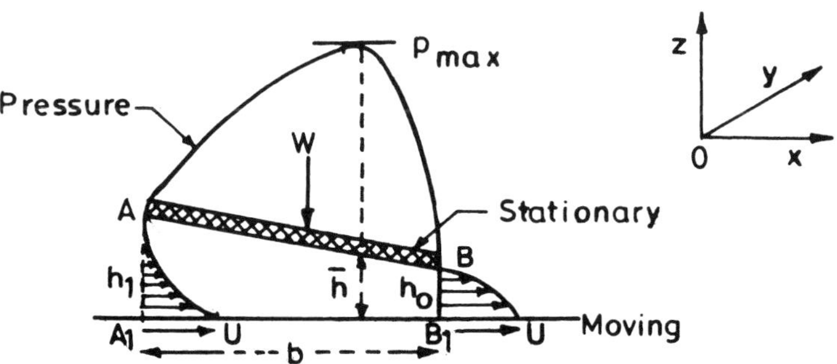

Fig. 1.7. Journal bearing showing eccentricity and pressure distribution.

where $\bar{h}$ is the film thickness at $\frac{dp}{dx} = 0$

The Eq. (1.27) can be easily solved analytically with the known boundary conditions at the inlet and outlet.

Side leakage cannot be neglected for finite sized bearings and the two dimensional equation has to be solved by finite difference methods. The Reynolds equation may be expressed in cylindrical coordinates in the case of journal bearings. With journal bearings, it is difficult to specify the boundary conditions in the divergent zone and some assumptions are necessary. Hamrock [26] has provided a detailed consideration of these methods. Graphical procedures are also available due to Raimondi and Boyd [27,28] that are commonly used. These procedures are also well discussed in [26]. The charts relate the bearing characteristic number to the minimum film thickness for different length/width ratios of the bearing. In the case of journal bearing the bearing characteristic number is defined as $\left(\frac{r}{c}\right)^2 \eta\left(\frac{N}{p}\right)$ where N is in revolutions per second and p is the pressure, N/m^2. Here p refers to the pressure based on the projected area. In the case of thrust bearing it is defined as $\left(\frac{1}{m^2}\right)\left(\frac{\eta u}{pb}\right)$ where $m = \frac{h_1 - h_2}{b}$ and u is sliding velocity, m/s. The film thickness increases with the bearing characteristic number. The temperature increases in the bearing due to friction and the effective film thickness has to be arrived at by an iterative procedure. Simplified graphical procedures are available to account for temperature rise and given in the earlier references [27,28]. The pressure effects on viscosity are not significant in hydrodynamic bearings and are neglected.

Bearings are designed so that the operating minimum film thickness is adequate for complete separation of the surfaces. The film thickness builds up from zero at the start to the designed value at the operating speed. Some boundary contact and wear are inevitable during start and stop of a bearing. The surfaces can be separated in the initial zone by an external pressure. Separation by external pressure is also possible between parallel surfaces. This is referred to as hydrostatic lubrication. As external pumping to high pressures is necessary with the associated hardware, this approach is not common.

1.6.3 Elastohydrodynamic lubrication

Concentrated stresses occur in line and point contacts. Such contacts occur in components like gears and rolling element bearings. Stresses and their distributions in line and point contacts are treated elsewhere in the book. For the present, only the available equations to calculate EHL film thickness are considered. At this stage, a brief consideration of the origin of EHL as a separate regime is in order. A rolling line contact involves equal velocities of the two rollers. A rolling/sliding contact involves unequal rolling velocities. In this case, the relative sliding velocity is the difference between the two rolling velocities. When the film thickness between two rolling/sliding cylinders was calculated by normal hydrodynamic theory the film thickness was found to be negligible. On the other hand, many gears that simulate such contacts were found to operate with negligible wear suggesting separating films. When pressure effect on viscosity and elastic deformation were taken into account, the film thickness calculated was found to be adequate to justify the observed operation in the gears. The final equations now commonly used are based on the complete numerical solutions as formulated by Dowson and Higginson [29] and Hamrock and Dowson [30]. The nature of the film and pressure distribution is shown in Fig. 1.8 for a line contact. It may be seen that the film is nearly parallel except for a small portion where there is a decrease in thickness. The point of minimum thickness corresponds to the pressure spike in the pressure distribution. The overall pressure distribution is similar to the normal pressure distribution based on Hertzian theory except for the pressure spike. The formula for minimum film thickness is given by

$$\frac{h_{min}}{R} = 2.65 \frac{G^{0.54} U^{0.7}}{W^{0.13}} \tag{1.28}$$

where

$G \quad = \xi E^*$ dimensionless material parameter

$U \quad = \eta_0 \dfrac{u_1 + u_2}{2E^* R}$, dimensionless speed parameter

$W \quad = \dfrac{w}{E^* R L_c}$, dimensionless load parameter

$R \quad = \left(\dfrac{1}{R_1} + \dfrac{1}{R_2}\right)^{-1}$, m

$E^* \quad = \left(\frac{1}{2} \left(\frac{1-\nu_1^2}{E_1} + \frac{1-\nu_2^2}{E_2} \right) \right)^{-1}$, N/m^2

w $\quad$ = total load in the contact, N

L_c $\quad$ = length of the contact, m

u_1, u_2 = velocities of the two surfaces

Equations are also available for point contacts. The relationships are similar except for some change in exponents.

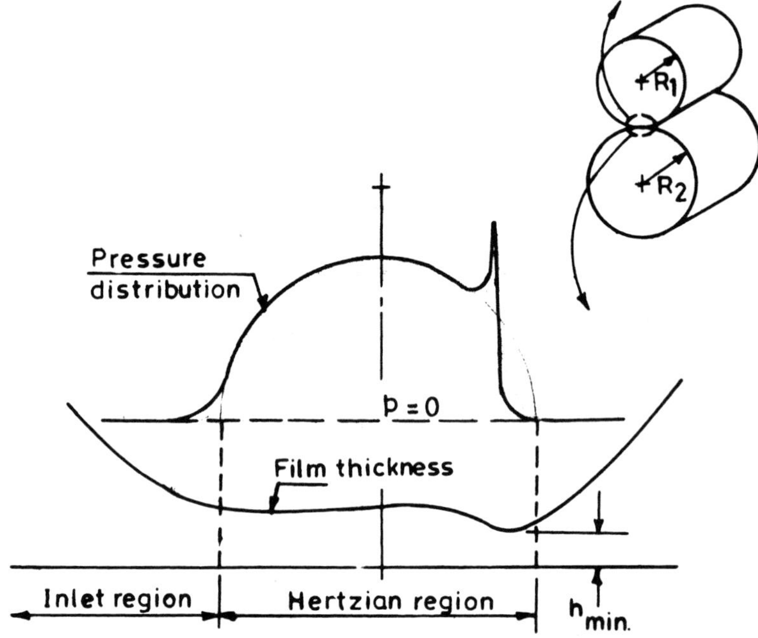

Fig. 1.8. Line contact illustrating film thickness and pressure distribution.

The above equation is for contacts in which pressure effect on viscosity and elastic deformation are significant as in the case of steels. Such a contact is designated as elastic-variable viscosity (EV) regime. Several other situations can occur in which these conditions are not applicable. For example in the case of a contact between steel and rubber the elastic deformation is dominant while the pressure effect on viscosity may be negligible. At the other extreme with two very rigid surfaces, the deformation is small while the pressure effect on viscosity is significant. Modified equations are available to treat such problems [30,31].

Film thicknesses in EHL range from 0.1 to 1.0 μm and are an order of magnitude lower than observed in hydrodynamic lubrication. Also, as can be seen from the equations load influence on film thickness is small. Rolling/sliding velocity, elastic modulus, and pressure coefficient of viscosity have a dominant influence on the film thickness.

1.6.4 Thinning films and boundary lubrication

It is desirable to operate tribological contacts with fluid films completely separating the surfaces. It is normally considered that complete separation occurs when the minimum film thickness is $3\sigma^*$ where σ^* is the composite roughness of the surfaces as defined in Eq. (1.10b). The ratio of film thickness to roughness is expressed as

$$\Lambda = \frac{h_{min}}{\sigma^*} \tag{1.29}$$

Many real components operate with Λ values less than 3.0 leading to asperity contacts. This can be easily appreciated with regard to EHL in which film thicknesses are of the same order as the roughness. Even with hydrodynamic conditions, the film thickness may reach very low values. Ring-liner contact in IC engine is a good example where the film thickness at the top and bottom dead centres is very low due to reversal of velocity at these points. At the middle of the stroke where maximum velocity occurs the film thickness is higher and complete separation of surfaces is possible. Even at the middle of the stroke the film thickness may be about 1.0 μm due to the complex nature of lubrication and high temperatures. As a rough guide, it is considered that when Λ is less than 0.5 the load is entirely supported by boundary layers. This regime is referred to as boundary lubrication. Boundary lubrication is mainly governed by the lubricant-solid interaction. Load is supported by both hydrodynamic and boundary films for Λ in the range of 0.5 to 3.0. This regime is referred to as mixed or partial hydrodynamic regime.

Film thickness is a function of operating conditions. With rapid technological changes, the equipment sizes are reducing and tribological components are operating under severe conditions. This leads to thinning films with increased boundary contact. Better methods of controlling wear and friction in boundary contacts are being increasingly sought. Another aspect of thinning films is the calculation of film thickness. Both EHL and hydrodynamic theories are based on smooth surfaces. When thin films are involved with Λ ratios of 6.0 or less the roughness influence on film thickness has to be taken into account. This is a difficult problem to be resolved. The problem is analogous to the flow of water in a channel. Flow of water can be easily modelled when the flow is normal. When the flow is reduced to a trickle one can observe meandering flow that is affected by boulders and the sand. In a similar fashion the flow, and hence the film thickness, are affected by roughness for low Λ values. It is also postulated theoretically that when the roughness has preferred orientation the film thickness is influenced positively or negatively by the lay direction. This idea opens up the possibility of "gaining" film thickness by manipulating the surface roughness. These issues will be considered in more detail in chapter 8 dealing with mixed lubrication.

References

1. D. Dowson, History of Tribology, Longman, New York, 1979.
2. F. P. Bowden and D. Tabor, The Friction and Lubrication of Solids, Part II, Oxford University Press, London, 1964, Chapter IV.
3. D. Tabor, Junction growth in metallic friction: the role of combined stresses and surface contamination, Proc. Roy. Soc. London, A 251 (1959) 378.
4. D. H. Buckley, Surface Effects in Adhesion, Friction, and Lubrication, Tribology series 5, Elsevier, Amsterdam, 1981.
5. P. M. Vedamanikam and D. V. Keller, A correlation between static adhesion data and the dynamic friction coefficients for two cobalt alloys and iron under vacuum conditions, ASLE Trans. 16 (1973) 73.
6. E. Rabinowicz, Friction and Wear of Materials, Wiley, New York, 1995.
7. N. P. Suh, Tribophysics, Prentice-Hall, New Jersey, 1986.
8. K. L. Johnson, Contact Mechanics, Cambridge University Press, London, 1985, 410.
9. J. A. Greenwood and J. P. B. Williamson, Contact of nominally flat surfaces, Proc. Roy. Soc. London, A 295, (1966), 300.
10. S. Ganti and B. Bhushan, Generalised fractal analysis and its applications to engineering surfaces, Wear, 180 (1995), 17.
11. D. J. Malvaney, D. E. Newland, and K. F. Gill, A complete description of surface texture profiles, Wear, 132 (1989) 173.
12. J. C. Jeagar, Moving sources of heat and the temperature at sliding contacts, Proc. Roy. Soc. NSW, 76 (1942) 203.
13. X. Tian and F. E. Kennedy, Contact surface temperature models for finite bodies in dry and lubricated sliding, J. Trib., ASME, 115 (1993) 411.

14. G. A. Berry and J. R. Barber, The division of frictional heat- A guide to the nature of sliding contact, J. Trib., ASME, 106 (1984) 405.
15. J. A. Greenwood, An interpolation formula for flash temperatures, Wear, 150 (1991) 153.
16. S. Jahanmir (ed.), Friction and Wear of Ceramics, Marcel Dekker, New York, 1992.
17. T. Saito, Y. Imada, and F. Honda, Chemical influence on wear of Si_3N_4 and hBN in water, Wear, 236 (1999) 153.
18. R. S. Gates, S. M. Hsu, and E. E. Klaus, Tribochemical mechanism of alumina with water, Trib. Trans STLE, 32 (1989) 357.
19. Y.Yamaguchi, Tribology of Plastic Materials, Tribology series, 16, Elsevier, Amsterdam, 1990.
20. G. M. Bartenev and V. V. Lavrentiev, Friction and Wear of Polymers, Elsevier, Amsterdam, 1981.
21. K. Friedrich (ed.), Advances in Composite Tribology, Composite Materials Series, 8, Elsevier, Amsterdam, 1993.
22. T. R. Choudhary, A. Sethuramiah, O. Prakash, and G. V. Rao, Development of polytetrafluoroethylene composite lining for a hydrogenerator thrust pad application, Proc. Instn. Mech. Engrs. 214, Part J, (2000) 375.
23. P. J. Blau, Friction Science and Technology, Marcel Dekker, New York, 1996.
24. C. M. Ettles and C. E. Hardie, The friction of some polymers and elastomers at high values of pressure X velocity, J. Trib., ASME, 110 (1988) 678.
25. C. J. A. Roelands, Correlation aspects of the viscosity-temperature-pressure relationship of lubricating oils, Druk, V. R. B. Groingen, Netherlands (1966).
26. B. J. Hamrock, Fundamentals of Fluid Film Lubrication, McGraw-Hill, New York, 1994.
27. A. A. Raimondi and J. Boyd, Applying bearing theory to the analysis and design of pad-type bearings, ASME Trans. 77 (3), 1955, 287.
28. A. A. Raimondi and J. Boyd, A solution for the finite journal bearing and its application to analysis and design (in three parts), ASLE Trans.1 (1958) 159.
29. D. Dowson and G. R. Higginson, Elastohydrodynamic Lubrication, Pergamon, Oxford, 1977.
30. B. J. Hamrock and D. Dowson, Ball Bearing Lubrication-The Elastohydrodynamics of Elliptical Contacts, Wiley, New York, 1981.
31. R. D. Arnell, P. B. Davies, J. Halling, and T. L. Whomes, Tribology Principles and Design Applications, Springer Verlag, New York, 1991.

Nomenclature

a' constant

A	dimensional constant
A_i	area of contact at asperity i
A_n	nominal contact area
A_r	total real contact area
b	slider bearing length
b'	constant
b^*	dimensional constant
c	radial clearance
c'	constant
c_p	specific heat
d	separation between rigid plane and mean surface height
E	effective modulus
E^*	effective modulus defined as E/2
E_1, E_2	elastic modulus of the two bodies in contact
f	coefficient of friction
F	friction force
G	dimensionless material parameter
H	hardness
$\overline{H}$	viscosity of the reference fluid with 100 VI in relation to test fluid
$\overline{h}$	film thickness at dp/dx = 0
h_1, h_2	inlet and outlet film thickness in slider bearing
h_{min}	minimum film thickness
k	thermal conductivity
k_I, k_{II}	thermal conductivities of the bodies I and II
k^*	$= \dfrac{k_I}{k_{II}}$
l	half contact dimension of square heat source
L	load in dry contact
L_c	length of EHL line contact
L_i	load on asperity i
$\overline{L}$	viscosity of the reference fluid with 0 VI in relation to test fluid
m	$= \dfrac{h_1 - h_2}{b}$
n	number of asperity contacts
N	revolutions per second
$\overline{N}$	$= (\log\overline{H} - \log U_t)/\log Y_t$
p	pressure

P_e Peclet number = $\frac{vl}{2\chi}$

q	heat flux
r	journal radius
R	equivalent radius of cylinders
R_1, R_2	radii of cylinders 1 and 2
R_a	centre line average surface roughness
R_q	root-mean-square (rms) surface roughness
s	shear strain rate du/dz
s_i	shear strength of the interface
s_m	shear strength of the metal
T	temperature
T_b	bulk temperature
T_c	contact temperature
u	sliding velocity
u_1, u_2	velocities of the two surfaces 1,2 in lubricated contacts
$\bar{u}$	mean surface velocity defined as $(u_1 + u_2)/2$
U	dimensionless speed parameter
U_t	viscosity of the test fluid at 40^0C
v	sliding velocity, m/s in dry contact
w	total load in EHL line contact
W	dimensionless load parameter
y	ordinate value in Fig. 1.5
Y_t	viscosity of the test fluid at 100^0C
z	asperity height over mean surface line

Greek Letters

α	constant
β	mean asperity radius
β^*	equivalent asperity radius
β_1, β_2	mean asperity radii of surfaces 1 and 2
χ	thermal diffusivity
δ_i	contact deformation of asperity i = $(z_i - d)$
$\Delta\theta$	temperature rise in the contact
$\Delta\theta_I$	surface temperature rise assuming all heat is flowing to body I
$\Delta\theta_{II}$	surface temperature rise assuming all heat is flowing to body II

ε	coefficient $= 0.398 y P_e^{-1}$
η	number of asperities per unit area
η	absolute viscosity
η_p	viscosity at pressure p
η_p	number of peaks per meter length
η_0	absolute viscosity at the given temperature and $p = 0$
φ	partition coefficient
$\varphi(z)$	height distribution function
κ	ratio of shear strength of the interface to the shear strength of the metal, $\frac{s_i}{s_m}$
Λ	ratio of film thickness to roughness
ν	kinematic viscosity
ν_1, ν_2	Poisson ratios of the bodies 1 and 2
σ	rms roughness of the surface
σ_1, σ_2	rms roughness of surfaces 1 and 2
σ^*	equivalent roughness
τ	shear stress in the fluid
ξ	pressure coefficient of viscosity at the given temperature
ψ	plasticity index

2. Lubricant Technology A Survey

2.1 Introduction

A lubricant may be defined as a solid or fluid film interposed between surfaces in relative motion to reduce friction and/or wear as defined in the previous chapter. The present chapter deals mainly with petroleum based lubricants used extensively in industry. Conventional and modern technologies involved in the manufacture of petroleum base stocks are considered first. Common additives and the formulation technology form the next section. Major lubricant specifications and test methods are then covered followed by a discussion of the issues related to performance. Tribological implications are also briefly considered and serve as a prelude to the main theme of this book. The next section deals with synthetic lubricants. The final section deals with the environmental issues involved in the use and disposal of lubricants. The literature available in lubricant technology is vast and the scope of the present chapter is limited to an appreciation of the modern technologies and their impact on lubricant performance.

2.2 Petroleum base stock manufacture

2.2.1 Crude distillation

Crude petroleum as received is first distilled in an atmospheric distillation column. Several side cuts are drawn from this column with varied boiling ranges representing gasoline, kerosene/jet fuel, diesel, and gas oil. Gaseous products are recovered from the top of the column. Gas oil may be subjected to other processes like fluidised catalytic cracking to obtain more valuable lighter products. In some cases crude may first be desalted as necessary. The product range suggested is typical but varies depending on the crude and market needs. The bottom product from this column, called reduced crude, is then subjected to vacuum distillation to obtain the base oils of different boiling ranges. While the term, 'base stock' is appropriate they are also called 'lubes' or 'lubricating oils' though they are not formulated lubricating oils. They are also referred to as mineral oils. The context in which these words are used is normally clear.

The basic scheme involved in distillation is shown in Fig. 2.1. Distillation is a complex process involving several side streams, several reflux streams, and steam stripping. In vacuum column steam is also used to reduce partial pressure. Crude consists of a large number of varied hydrocarbons and empirical procedures were used in the design [1]. The more recent approaches consist of dividing the crude into several close boiling components, called pseudo components, and then using elaborate computer programs for design [2]. Such procedures, which approach rigorous multi-component distillation column design, provide better control of the distillation process. Referring to Fig. 2.1 the crude is first heated in a furnace and then fed to the atmospheric column where it is distilled. As shown, several side streams are withdrawn after steam stripping. Reflux streams can also be seen including the top reflux. Uncondensed gases are recovered from the top reflux drum and used for further processing. Vacuum distillation can yield several cuts called neutrals and in the present scheme three side draws are shown. The top product can be a light viscosity cut termed as spindle oil. The bottom product from vacuum column is called vacuum residue and in some cases can yield very heavy oils called bright stocks by deasphalting. It may be noted here that all reduced crudes are not suitable for lube manufacture and must have high molecular weight components of desirable structure. Vacuum distillation can also be practiced on the reduced crude, which is not lube bearing. The distilled high boiling fractions, called vacuum gas oils, can be cracked to obtain desirable fuels. The viscosities of base oils produced in a lube distillation unit can range from 10 cSt at 40°C for spindle oils to as high as 600 cSt at 40°C for bright stock with intermediate values for the neutrals. Neutral is normally preceded by a number, which refers to the older Saybolt Universal Seconds (SUS) viscosity based on time to flow through an orifice. With regard to heavy bright stocks, the convention is to precede the SSU value at 210°F. The equivalent cSt values at 40°C for neutrals approximately range 0.19 to 0.21 times the SSU value at 100°F. The fuels as well as lubes undergo further refining of varied levels. This is necessary to meet the specifications of these products.

Though the main interest here is with regard to lubes it is of interest to consider briefly the tribological problems related to fuels. In modern refineries aviation turbine fuel (ATF), also called jet fuel, is subjected to hydroprocessing to improve quality. The processing removes most of the polar impurities. As the fuel itself lubricates piston pumps that deliver fuel to the combustion chamber, the better fuels are found to be deficient in boundary lubrication leading to wear and seizure problems. The associated problems are well documented [3,4]. A recent test method, ASTM D5001, is aimed at assessing the lubricity of ATF by a ball-on-disc laboratory rig. Another example is the diesel fuel now being manufactured with very low sulphur, which is causing wear problems that are currently under investigation [5,6]. Boundary lubrication problems with fuels are thus also considered as lubricated wear problems.

2.2.2 Refining of base stocks

The base oils obtained by distillation need further refining to improve their characteristics. The selected processing depends on the nature of the base stock and the desired quality of the refined base oils. It is hence necessary to appreciate the nature of base oils as obtained by distillation and then go on to a consideration of refining processes.

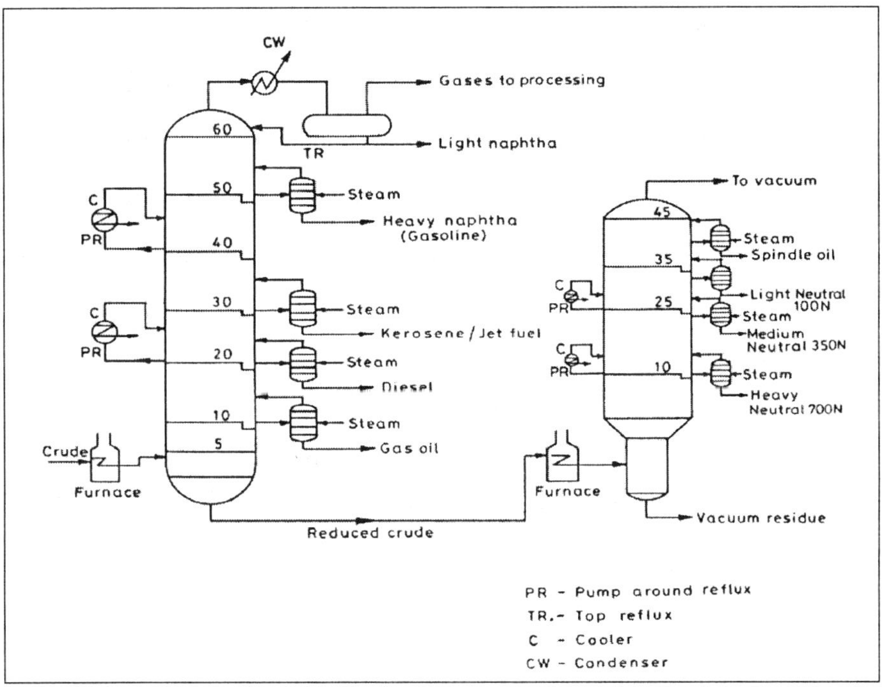

Fig. 2.1. Crude distillation process. Plate numbers are indicative only.

2.2.3 Base oils

Crude oils are classified as paraffinic, napthenic, and aromatic depending on the predominant structures in the complex molecules. Base oils obtained by distillation are similarly termed and correspond to the original crude. Aromatic crudes are not used for the manufacture of base oils as aromatics have poor viscosity-temperature relations and decrease rapidly in viscosity with temperature. They also have thermal and oxidative stability problems. Paraffinic structures have very good viscosity- temperature relationship besides stability. Thus, most of the base oils manufactured in modern refineries are paraffinic in nature and the focus is on these oils. Naphthenic oils, which are intermediate in nature, are manufactured in limited quantities, mainly for low temperature applications because of their good low temperature fluidity.

The paraffinic base oils range in molecular weight from C_{25} to C_{40}. Most of the molecules in the oils will have mixed structures. For example a long paraffinic side chain may be attached to a benzene ring. The structure of this molecule is mainly paraffinic in the sense that most of the carbon atoms in the molecule are of paraffinic nature. To appreciate the structures better Fig. 2.2 illustrates three typical molecules in which the normal and isoparaffin chains are linked to a saturated ring, condensed naphtheno-aromatic rings, and to a 3-ring aromatic structure. The structure (a) will have high paraffinic nature followed by (b) and (c). The structure (c) is useful only if the aromatics are saturated. It may be appreciated that several other combinations are possible. It is apparent that paraffinicity is relative. There are two approaches available to characterise the nature of hydrocarbons involved. One is the difficult process of isolating the molecules to the extent possible by a combination of gas chromatography and mass spectrometry and determining the structures. Nuclear magnetic resonance spectroscopy is also used to understand the structures in more detail. Dorinson and Ludema [7] have reported a good survey of these methods. The other approach is to estimate the paraffinic, napthenic, and aromatic carbon content by indirect methods like n-d-M (n is refractive index, d is density and M is molecular weight), characterisation factor K, and other techniques which are well documented by American Petroleum Institute [8]. These methods are based on the fact that density, boiling range, and refractive index vary with the nature of hydrocarbons and hence can be correlated with the overall nature of the fractions involved. These approaches, which are indirect, have stood the test of time and are extensively used. These techniques may not be adequate to characterise very heavy crudes, which the refiners are increasingly obliged to process. Also more advanced methods are needed to characterise the polynuclear aromatics. These issues are well covered in a recent book [9]. The other structures in the base oils are those containing hetero-atoms, which are mainly sulphur and nitrogen. Some oils may have significant sulphur and nitrogen, which need to be removed by hydrogenation.

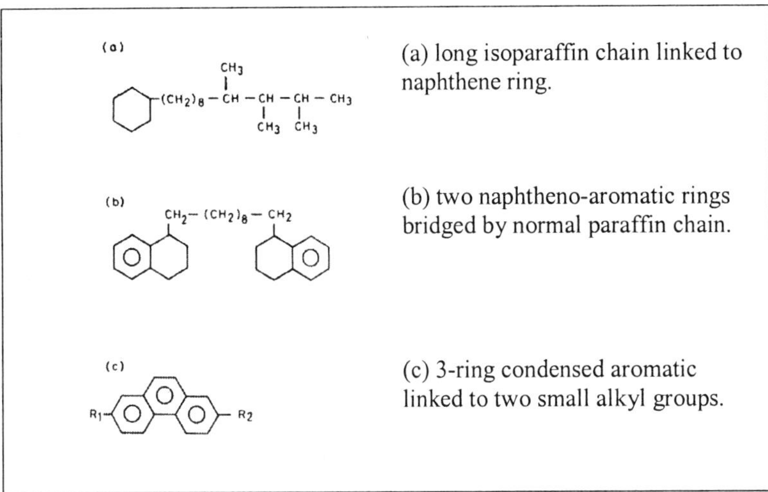

Fig. 2.2. Typical structures in base oil.

2.2.4 Technology of refining base oils

The distilled paraffinic oils need upgrading in three areas. One is the reduction in aromatic content to improve VI and oxidation stability, second is the removal of long chain normal paraffinic structures which form wax and create low temperature flow problems, and the third is the removal of sulphur, nitrogen and other polar impurities to acceptable levels. The third is also referred to as a finishing process. It is also termed hydrofinishing when it involves hydrogen treatment. Aromatic removal can be affected by solvent treatment. The main solvents used in the industry for this purpose are furfural, N-Methyl-2-Pyrrolidone (MP), and phenol. These solvents preferentially dissolve the aromatic constituents thereby enriching the raffinate with desirable constituents. The extract is stripped of the solvent and the aromatics are utilised as feed stocks for petrochemical industries or fluidised catalytic cracking. Instead of removing, the aromatics can be converted to saturated cyclic structures by hydrogenation. Hydrogenation also simultaneously removes sulphur and nitrogen. Solvent extraction involves loss of product while there is no loss in hydrogenation as aromatics are converted and retained. Solvent extraction can also be done on the vacuum residue to obtain very heavy oil called bright stock. The solvent used for this purpose is propane, which preferentially dissolves the oil fraction and rejects the asphaltenes. This process, called propane deasphalting, is done only if the bright stock content is high enough to economically process the residue. The conventional dewaxing process is again solvent based. The major solvent used is methyl ethyl ketone (MEK). The wax is precipitated from the solution by chilling to low temperature and then filtering. The hydrofinishing is meant to remove polar impurities by a hydrogenation process and is the final step to improve colour and stability. This brief discussion suggests that in a refinery several combinations of treatment are possible which indeed is the case. The flow diagram given in Fig. 2.3 illustrates the various combinations possible. Chemical processes consisting of acid and alkali treatments are normally not used in modern refineries though mentioned here. Catalytic dewaxing mentioned is a modern process which is considered next.

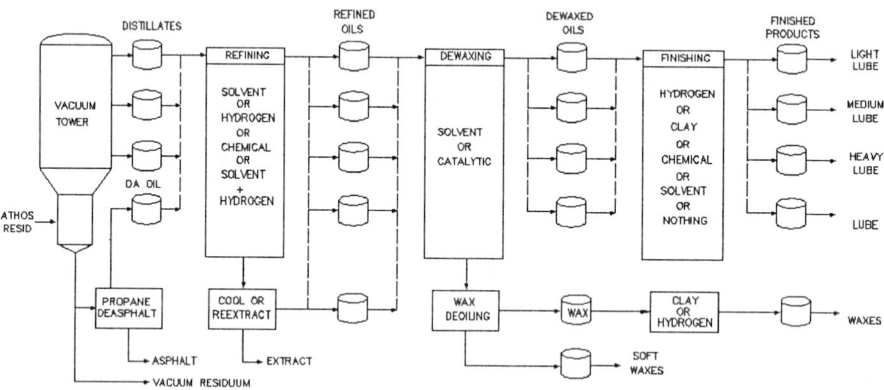

Fig. 2.3. Refining of base oils. (From Ref. [11] p2, by courtesy of Marcel Dekker Inc.)

The modern refining processes include catalytic dewaxing, catalytic isomerisation, and hydrocracking. Catalytic dewaxing and isomerisation processes involve selective zeolites with specific pore sizes that preferentially allow long chain paraffins into the pores where they undergo transformations. These transformations can involve cracking leading to lighter ends, which can be distilled off to obtain fuels, or moderate cracking coupled with isomerisation resulting in isoparaffins that have high VI. These processes are conducted under hydrogen pressure leading to saturation of the molecules. These processes are very specialised and are made possible by the major developments in zeolite manufacture leading to controlled pore sizes. Different metals are incorporated to influence the hydrogenation function.

Hydrocracking is a severe cracking and hydrogenation process originally developed to manufacture high quality fuels. The process is flexible and, for example, typical heavy gas oil can be cracked to diesel or gasoline preferentially by changing the operating conditions in a hydrocracker. With developments in catalysts including higher tolerance to impurities, the process in the recent decade is being adapted for lube manufacture.

The flexibility is such that vacuum gas oils which are not lube bearing in the conventional sense can be converted to base oils of high quality. Hydrocracking also removes sulphur and nitrogen during processing. The reactions involved in hydrogenation are shown in Fig. 2.4. Besides saturation, hydrocracking also involves isomerization and the opening up of naphthenic rings resulting in paraffinic structures. The catalyst has cracking and hydrogenation functions, which can be selectively controlled, by the extent of zeolite in the silica-alumina base and the type and concentrations of metals used. The metals used include cobalt, molybdenum, nickel, and tungsten. Noble metals like platinum are also incorporated in some cases. A general coverage of hydrogen processing and catalysts is available in a recent handbook [10] while specific consideration with regard to lubricants is available in the book by Sequeira [11]. An example can explain the flexibility involved. Consider a highly aromatic high boiling component unsuitable for lubricant manufacture in the conventional sense. This component can undergo significant saturation via hydrogenation. The naphthenic structures can further undergo ring splitting leading to desirable paraffinic structures resulting in good base oil. In the case of long chain waxy feed, waxy structures can be partially cracked and hydrogenated leading to less viscous, but more acceptable, base oils. Thus, there are several combinations, which can be adopted and integrated into overall lube manufacturing schemes. It needs to be pointed out that all treatments involving hydrogen may be categorised under hydroprocessing. The difference between the various categories is the relative emphasis on cracking and hydrogenation. The operating conditions, catalysts, and hydrogen consumption vary with the nature of treatment. Thus hydrogenation used in place of solvent treatment to saturate aromatics involves negligible cracking while hydrofinishing involves final removal of polar impurities and involves a mild treatment. The wax cracking mentioned above is also a hydrocracking process, but it is selective to waxes.

It is here necessary to briefly consider other issues involved in hydroprocessing. The poisoning of catalyst limits the efficiency and cycle life of a catalyst and in some cases pretreatment of feed is necessary to reduce impurities. Reaction rates for various components can be vastly different. As an example, desulphurisation of thiophene can occur 27 times faster than dibenzothiophene [12]. Hence major structures involved in the feed and their reaction kinetics are to be understood in selecting the nature of the process and the operating conditions. The costs involved in severe processes like hydrocracking can be significant in terms of investment and operating expenses. Hydrogen availability is another major factor. These aspects deter some of the refiners from adopting large scale hydroprocessing.

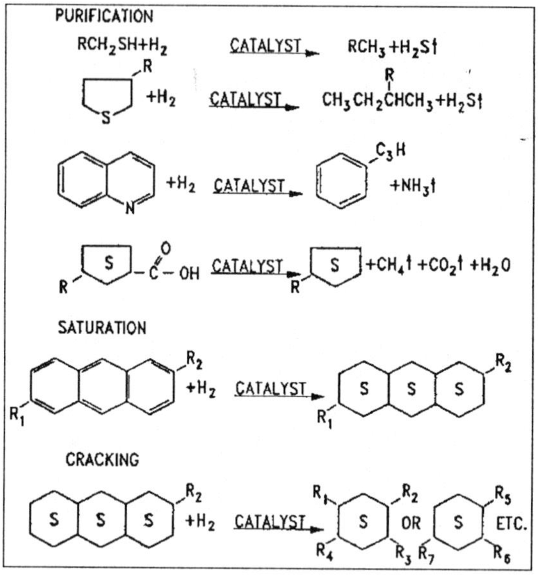

Fig. 2.4. Chemical reactions in hydroprocessing. 'S' refers to sulphur in the first reaction and to saturated rings in others. (From Ref. [11] p120, by courtesy of Marcel Dekker Inc.)

2.3 Additives and formulation technology

Finished base oils as obtained from refinery are normally blended with additives to obtain lubricating oils used in industry. Additives are added to enhance the existing properties of the base oil and to impart new specific properties.

2.3.1 The nature of additives

It is convenient first to appreciate the overall physicochemical properties of lubricants as given in Table 2.1. The test methods are given in ASTM standards for petroleum products [13]. While these are American standards, many other standards are available from organisations of different countries, which in some cases differ from ASTM standards. These standards detail the procedure required to conduct the tests and include information on repeatability and reproducibility of each test. The table also gives the central purpose of each test. The lubricant specifications prescribe the acceptable values of the various properties and are drawn up by different organisations in various countries. There are thus a large number of specifications for lubricants. The original properties of the base oil strongly influence the additive level to meet the required enhancement. Another important aspect of additives is that more than one additive is routinely added to improve several properties together. It is important that such systems do not interact with each other in an antagonistic manner. Sometimes the additives may interact synergetically enhancing the effect of each, which is desirable. When the additive system does not interact negatively it is said to be compatible. Yet another aspect is the ability of different commercial oils for the same end use to be compatible. As an example, if a customer is obliged to mix different brands of engine oils they should not adversely affect the performance. The major categories of chemical additives used for enhancement of properties are described below. The detailed treatment of chemical structures is available in the literature [14,15]. Lubricants and their applications are covered in several books [16,17,18] and the following is a limited treatment. Detergent additives consist mainly of oil soluble sulphonates and phenates of calcium, barium, and magnesium. These additives have a major role in engine oils to prevent deposition of carbon and varnish in the piston-ring zone. Modern detergents are overbased to provide a reserve of alkali to neutralise the acidic components formed during combustion. Dispersants again are mainly used in engine oils to prevent low temperature sludge deposition and include succinimides, polyamides, and copolymers containing polar groups. The antioxidants used include hindered phenols like ditertiarybutyl paracresol and amines like β-naphthyl amine. These additives find application in turbine and hydraulic oils. Zinc dialkyl dithiophosphates are a very special class of additives extensively used in engine oils. These additives are multifunctional and can simultaneously function as antiwear, antioxidant, and anticorrosive additives. Silicones are the usual antifoam agents used at ppm level to control foaming in turbine oil, hydraulic oils, and gear oils when necessary. The VI improvers mainly consist of long chain polymers, which have higher solubility at higher temperatures thereby compensating for the loss of viscosity of the base oil. This results in higher VI for the system. The major categories used are polymethacrylates, polyisobutenes, and olefin copolymers. The molecular weight of the polymers used is above 50,000. VI improvers are mainly used in engine oils and also in some other applications like gear oils. Pour point depressants are polymeric molecules

Table 2.1: Physicochemical properties of lubricants

Property	Test Method (ASTM)*	**Comments**
1. Flash and fire point	D92, D93, D56, D130	Approximate characterisation of flammability. Specification control and for detection of mix-up with higher/lower boiling components.
2. Aniline point	D611	Indicates aromaticity and useful to check seal compatibility.
3. Neutralization number Total acid number Total base number	D974, D664	Specification control on base stocks and finished products. Increase in acidity is generally an indication of oxidative degradation oil. Base number indicates alkali reserve in engine oils due to detergents.
4. Water content	D95, D1744	Water deleterious with regard to oxidation and corrosion. Control and estimation critical in turbine oils, hydraulic oils, and transformer oils.
5. Carbon residue	D189, D524	Indicates tendency for carbon deposits. Less widely used now.
6. Pour point and cloud point	D97, D2500	Relevant with regard to low temperature flow properties within limits. More representative methods utilising rotary viscometers are now being used to characterise flow.
7. Ash content, sulphated ash content	D482, D874	Indicates overall metallic content due to additives. Increase in use indicative of wear debris accumulation.
8. Sulphur content	D129	Sulphur in base oil limited by specification. Also estimation gives sulphur content of additives when present.
9. Copper strip corrosion	D1275	Indicative of 'active' sulphur content. Important with regard to copper containing metals like bronzes. Also indicative of refining level of base stocks.
10. Viscosity and viscosity index	D445, D2270	Major factor in lubricant selection and invariably specified. Increase in viscosity in use is indicative of oxidation.
11. Oxidation stability	D943 and other tests	Usually long duration tests. Critical for applications like turbine and hydraulic oils.
12. Demulisibility characteristics	D1401	Indicates the ability to separate water and is important for turbine oil, hydraulic oil.
13. Foaming characteristics	D892	Foaming means ineffective lubrication and should be controlled.
14. Additive elements	Modern methods like emission spectroscopy or wet methods listed in ASTM.	Very important to know the level of additives and how they are depleting.
15. Tribological tests	D2783, D4172 and others	Estimation of wear and load bearing capacity at a given set of conditions.

*ASTM methods are representative only. A large number of methods are available in some cases like oxidation tests.

with a polar group. These molecules prevent agglomeration of wax crystals and improve the low temperature flow properties of the lubricant. These additives again have major application for engine oils. Extreme pressure (EP), and antiwear additives are sulphur-phosphorous compounds used to control wear and scuffing and are dealt with in the chapters on boundary lubrication and lubricated wear.

2.3.2 Formulation methodology

Base oils with the required characteristics are blended with the selected additive package. The modern blending plants are automated with precise control of the additive dosage and blending parameters. The product is then evaluated as per specifications set for the finished lubricant. If the lubricant does not pass the specifications, the formulations are re-worked until the specifications are met. For formulations already set, the procedure involves routine manufacture. The additives may be available as a package from additive manufacturer or may be individually purchased, blended, and optimised by the formulator. The lubricant formulation is no longer a simple blending of known packages. It is now a very dynamic and demanding technology. The first factor is the continuing demand from equipment manufacturers for better products with longer life under increasingly severe operating conditions. These demands have their effect on specifications, which get more stringent. In some cases the equipment manufacturer may demand qualification against some in-house tests in addition to the specifications. The other factor is the increased environmental protection demanded of the lubricants, which translates into regulations on toxicity and bio-degradability. In addition, the formulation will have to be cost effective to compete in the market. These demands also have their influence on the quality of base oils. This is indeed the reason for increasing levels of hydroprocessing resulting in more stable and very high VI base oils. Earlier base stocks were considered adequate with 95 to 100 VI. Today oils with VI above 120 have become common. These hydroprocessed oils have also some deficiencies like poorer solubility for additives. These are taken care of by modification of additives. The demands also have an impact on additive technology, resulting in better additives. The methodology available for performance testing is a major problem. Laboratory tests in many cases cannot correlate well with practice, necessitating expensive field trials. With regard to engine oils, several engine tests may be needed for qualification which is an expensive and time consuming process. This important issue of correlation between laboratory and field performance shall be considered later.

2.4 Lubricant specifications

Lubricant specifications define firstly the limits on physicochemical characteristics, which are appropriate for the particular lubricant. They also specify limits on performance in laboratory rigs as applicable. The term laboratory rig is used here with reference to mechanical test rigs. It is common to divide lubricants into automotive and industrial lubricants. Automotive lubricants consist of engine and gear oils. Engine oils form about 50% of the lubricants manufactured and have extensive engine tests incorporated in their specifications. Thus they need to be treated as a separate class.

2.4.1 Automotive lubricants

Engine oils are the major category of automotive lubricants. The first classification to be considered for engine oils is the well-known SAE viscosity classification. This classification is based on the viscosity ranges of the lubricants in cSt at 100°C. In all six winter grades and six summer grades are specified. The winter grades range from 0W to 25W while the summer grades range from 20 to 60. The recent SAE specification SAE J300-1999 specifies important additional requirements. It specifies low temperature cranking and pumping requirements for the winter grades which must be additionally met. The low temperature rheological behaviour of lubricants is complex and the prescribed tests conducted with rotary viscometers provides reasonable assurance of field performance. Another important development is the high temperature high shear (HTHS) viscosity at 150°C specified for the summer grades. This is because the recent research has shown this viscosity is more relevant with regard to bearing performance.

It is necessary here to consider the concept of multigrade oil. This refers to those oils which can meet the winter and the summer grade simultaneously. Thus 20W-40 means oil that meets 20W as well as 40 grade. This is made possible by adding VI improvers to the relatively low viscosity winter grade oil. While the original purpose was mainly to avoid two separate grades, the multigrade oils are capable of reducing energy losses due to lower friction. The stability of VI improvers at high shear rate in the ring-liner zone and their rheological behaviour are now areas of major investigation. Several laboratory tests are available to study the shear stability of VI improvers.

The performance with regard to detergency, sludge, corrosion, and wear is necessarily determined by special engine tests. This is because effective laboratory small scale tests are at present not available for such evaluations, though several methods have been attempted. Detailed consideration of the engine tests is not within the scope of the present chapter and only the major approaches involved are indicated. Diesel and gasoline engine oils are classified into several categories depending on the severity of operation. For each category, several engine tests are specified. The main aspects studied are oil oxidation and bearing corrosion, rust, high temperature deposits valve train wear, and low temperature sludge. Some engine tests are also conducted to study ring-liner wear and fuel economy. Several organisations are involved in drawing up classification systems and engine tests. These include the American Petroleum Institute, US Military, and European Commission. The engine tests are changing continuously and it is common to observe that several tests become obsolete and are replaced by new ones. At present there is no other approach available to ensure lubricant performance. The severe performance requirements are being met by substantial improvements in additive technology and base stocks. The base oils now available are categorised into five types as per API 1509. Types I, II, and III are petroleum based while IV and V are synthetic based. Type III is a very high VI base stock with a minimum VI of 120 and is now replacing other types, mainly in engine oils. High to very high VI base oils are now available because of the modern hydroprocessing technologies discussed earlier.

The detailed engine test procedures are available through the standards organisations. A list of engine tests currently used by API to qualify diesel engine oils is given in Table 2.2. This is to illustrate the complex testing requirements. Each lubricant category will have its specific engine test requirements. Thus API CF-4 will have to pass Caterpillar 1K, Mack T-6, and Mack T-7 tests to qualify.

Two-stroke engine oils have special requirements and are designated by API into five categories TA to TE with increasing severity. The engine tests used evaluate lubricity, ring sticking, detergency, and the tendency for pre-ignition. The engine tests involved are jointly developed by CEC, ASTM, API, and SAE. 2–Stroke engines are extensively used in India and China and there is special interest in developing specifications in this area. The Indian standard IS: 14234-1996 covers engine tests based on indigenous 2-Stroke engines. Japan also has detailed engine tests based on their JASO standards. Rigorous specifications are also to be met by the automatic transmission fluids and gear and axle oils.

The viscosity grades for gear oils are governed by SAE J306 recommended practice. These cover nine grades ranging from 70W to 250 with corresponding

Table 2.2: Current diesel engine tests for API category lubricants.
(Source: lubrizol.com/reference library)

Parameter	Caterpillar 1k	Caterpillar 1M-PC	Caterpillar 1N	DDC 6V92TA	Mack T-6	Mack T-7	Mack T-8	Roller-Follower Wear
Engine Displacement (cu cm)	2400	2000	2400	9046	11,000	11,000	11,930	6200
Test Duration (h)	252	120	252	100	600	150	250	50
Test Phases	1	1	1	2	3	1	1	1
Oil Change Period (h)	None	None	None	None	None	None	None	None
Engine Speed (rpm)	2100	1800	2100	1200 & 2300	1400-2100	1200	1800	1000
Fuel Rate (kW)	140	103	140	653 & 1130	523-697	514	795	113
Overall Load (kPa)	1240	972	1240	945 & 1075	1262-1531	1655	1458	636
Temperature (°C)	127	124	127	103 & 143	66-99	57.2	43	Not applicable
Absolute Pressure (kPa)	240	179	179	248 & 352	98-155	101	203	Not applicable
Water Outlet Temperature (°C)	93.3	85	93.3	84	87.8	85	85	118
Oil-to-Bearing Temperature (°C)	107.2	96.1	96.1	102.2 & 111.1	112.8	112.8	104.4	118
Fuel Sulphur (% mass)	0.38-0.42*	0.35	0.05	0.1-0.4	0.1-0.3	0.40 max	0.03-0.05	0.03-0.05
Test included in specifications	API CF-4	API CF, CF-2, MIL-PRF-2104G	API CG-4, MIL-PRF-2104G	API CF-2, MIL-PRF-2104G	API CF-4	API CF-4	API CG-4, MIL-PRF-2104G	API CG-4, MIL-PRF-2104G
Test evaluates	Piston deposits, oil consumption	Piston deposits, wear	Piston deposits, oil consumption	Piston, liner, and ring distress	Piston deposits, wear, oil consumption, oil thickening	Oil thickening	Oil thickening	Lifter pin wear

viscosities of 4.1 to 41.0 cSt at 40°C. The conditions are particularly severe for hypoid axles and tests are required on specified axles. Two important tests are CRC L-37, which is a low speed high torque test, and CRC L-42, which is a high speed shock loading test. These tests can be met only with a high level of extreme pressure additives containing organic sulphur and phosphorous compounds. This is a classic example of successful use of additives. The operation of hypoid axles without scuffing is not possible without these additives. The other tests include oxidation stability, rust, and foaming charecteristics. The API designation GL-5 covers these requirements for severe applications. The manual gearboxes may be lubricated by lower grades like GL-4 or GL-3. Automatic transmission fluids are now highly specialised and have to meet various requirements. Besides physicochemical tests, wear tests on laboratory rigs are specified. Friction durability tests and shudder tests are also carried out. The shudder test is intended to ensure freedom from stick-slip behaviour. The specifications currently used are from Ford and General Motors of US and are designated by MECRON-V and DEXRON-III. These specifications are also followed in Europe and Japan with some modifications.

2.4.2 Industrial lubricants

The major classes of industrial lubricants may be divided into metal working fluids, hydraulic oils, turbine oils, and gear oils. Many applications for industrial lubricants are less severe than for engine oils, but laying down specifications for industrial oils has its difficulties of a different kind. These difficulties arise from very diverse applications of these products in a given category and it is impossible to prescribe tribological rig tests in each case. Thus, metal removal processes can involve cutting, drilling, tapping, honing, grinding, and in each operation, again there can be wide variations. The specifications have to necessarily end up with a compromise, which is a minimum requirement. Such compromises are inadequate to predict performance effectively and reliance has to be placed on the knowledge of the lubricant supplier and the equipment manufacturer, particularly for severe operations. For many new applications assured performance can be based on field trials only. This is unlike the case for engine oils where more reliance can be placed on the engine tests for selecting formulations. The specifications considered below should be viewed within these limitations.

2.4.2.1 Metal working fluids

Metal working fluids are divided into water based emulsions and neat fluids. The water based fluids consist of oil in water emulsions and are widely used in metal cutting operations. Besides emulsifiers several other additives are added depending on the conditions of operation. Antifungal and antimicrobial additives are common to prevent bacterial growth in aqueous systems that can cause skin dermatitis. Mild EP additives are also added for more severe operating conditions. Neat fluids, also called oil-based fluids, have requirements for EP additives, and demulsifiers to readily separate water. Oil mist is a problem in these operations, which is taken care of by antimisting agents. The specifications closely control the corrosion, rusting, and oxidation by laboratory tests. The tribological tests prescribed include the well-known 4-Ball EP and antiwear tests, Timken OK load test, and Falex EP test. These tests are governed by ASTM D2783, D2266, and D3233 respectively. Depending on the application, some of these tests and the limits are specified. However in view of their inadequacy many organisations rely on in-house tests. The Falex torque test involving measurement of torque during tapping is an additional test used by some organisations. Metal forming lubricants are a special class and dealt quite extensively in two books [19,20]. The major problem in metal forming is to ensure tribolgical performance. Normal tribological tests mentioned above are used to screen products. These tests do not involve gross plastic deformation and several tests involving plastic deformation have been developed. These methods have limited applicability, as they cannot simulate varied metal forming operations.

2.4.2.2 Hydraulic oils

Most of the modern hydraulic oils are antiwear oils containing mild EP additives. They also have rust and oxidation inhibitors incorporated into the formulation. Hydraulic oils are expected to last for a long duration and so an oxidation test, ASTM D943 is incorporated. Oxygen is bubbled at controlled temperature in the presence of water and catalysts in this long duration test. Acidity level after 1500 hours operation forms the test criterion. In the case of turbine oils, which shall be discussed next, the acidity is tested after 2000 hours. The other tests include demulsibility to ensure effective separation of water. The performance test is normally conducted in piston or vane pumps and limits are laid down. The major specifications are due to Denison, Vickers, and Cincinnati Milacron. It is of interest to note that wear testing is conducted directly on the pumps used in the hydraulic systems and the evaluations are expected to be more reliable. The oils are also evaluated for air release property by ASTM D3427. The ability to quickly release air from the system is important to operate the pumps without encountering the cavitation problem.

2.4.2.3 Turbine oils

Turbine oils are unique for their long life with about 10% topping per year. Many steam turbines can operate for ten years without any oil change. Even gas turbines can operate for three years without oil change. This is achieved by special base stocks that have excellent water separability and good oxidation stability. Turbine oils are divided into R&O and EP categories. The R&O category refers to the rust and oxidation inhibited oil. It is fortified mainly with antioxidant and rust inhibitors along with a foam inhibitor. Such oils are used for systems that do not involve reduction gears. In cases where reduction gears are involved, turbine oils with mild EP additives are used. When EP additives are involved a tribological test in the FZG gear test as per DIN 5134 is specified. Another oxidation test, called rotary bomb oxidation test (RBOT) as per ASTM D2272 is also sometimes specified. RBOT is more often used to study residual life of turbine oils. It may be noted that several viscosity grades are available for turbine and hydraulic oils and are normally specified as per the ISO classification. Air release is also specified, as in the case of hydraulic oils.

2.4.2.4 Industrial gear oils

The widely used specifications for gear oils are those specified by the American Gear Manufacturers Association (AGMA). The gear oils are divided into nine categories ranging from 5 to 50 cSt in viscosity at 40°C. The corresponding ISO grades range from 48 to 1000. The other known specifications are due to US Steel and David Brown. The tests include oxidation stability, rust, foaming, and demulsibility. Tribological tests are included for EP grade oils. The AGMA specifies FZG and Timken OK load tests. US Steel includes also the 4-Ball EP and antiwear tests. In view of the large varieties of gears used field performance cannot be correlated with these rig tests. The tests are, however, useful in ensuring that the lubricants have adequate EP and antiwear levels at least in the prescribed rigs. The FZG test is considered by industry as more reliable than the 4-Ball EP test.

2.5 Performance issues

Lubricant technology has advanced to a high level of sophistication as already noted. High quality base oils obtained via hydrogenation and corresponding developments in additives are helping to meet stringent specifications. Synthetic fluids are competing for a place but they still form a minor share of the market. The present section concerns itself with petroleum based lubricants. Despite careful formulation, problems do arise in use. Bore polishing problems a decade back and the more recent problem of black sludge in engines are examples of such problems. Such problems are solved as they arise with cooperative effort between equipment manufacturers, oil companies, and additive manufacturers. One serious gap that remains is the inability to predict performance based on laboratory tests except in some cases. This has implications for the quicker development of products, as at present candidate products cannot be screened effectively in a shorter time. In addition, it has implications with regard to customers consisting of both equipment manufacturers and end users. Customers often want to select amongst competing products all of which meet a given specification. Independent laboratories can do this provided there are methodologies available. Another area of customer interest is to know the rejection limits in service. While great strides have been made in additive development and selection of base oils scientific understanding of performance has lagged behind. The present section deals with the broader issues of methodology. The problems involved are first illustrated with the example of turbine oil which is a relatively simple formulation. The issues related to engine oil performance are then considered. This is followed by a consideration of the tribological problems related to lubricants.

2.5.1 Oxidation of turbine oils

Turbine oils are evaluated to ensure oxidation stability by ASTM D943, in which oxidation is conducted at 95°C in the presence of copper and iron catalysts and water. Oxygen is bubbled through the oil at a prescribed flow rate. The acidity developed over a long duration of 2000 hours forms the criterion and is normally limited to 2.0 mg equivalent of K(OH). Excellent water separabilty and good air release properties are also required. Even for this simple system, the oxidation stability cannot be predicted based on the composition of the base oil. The main problem is in understanding the influence of different constituents in the base oil. It is known that polynuclear aromatic structures, as well as complex sulphur and nitrogen compounds, have a significant influence and these compounds cannot be characterised in detail. Also the combined effect of these compounds is difficult to predict. The complex problem has recently been studied by neural network simulations, which related oxidation to seven likely constituents of the base oil with a given additive system and different base stocks [21]. The purpose in citing this example is to show that knowledge levels are not adequate for the prediction of oxidation even for simple systems. The 'vagaries' of the base stock are likely to be less for the hydroprocessed base stocks of type III described earlier which contain very low aromatics, sulphur, and nitrogen. It is now demonstrated that such oils, with the usual levels of additives, can have oxidation life 2 to 3 times higher than the conventional oil formulations [22].

Fundamental oxidation studies on hydrocarbons are well documented and Igarashi et al [23,24] have used the available information on rate constants to develop an effective computer simulation model to predict oxidation behaviour of hexadecane in the presence of di-tertiarybutyl paracresol (DBPC), which is a commonly used antioxidant in turbine oils. Their model also took into account the direct oxidation of DBPC which is an important consideration. They then tried to treat turbine oil oxidation in terms of a mixture of hexadecane and tetralin, which are well-defined hydrocarbons amenable to kinetic treatment. Such extension to turbine oils is open to debate, but the oxidation modelling for pure hydrocarbons may be adaptable to pure synthetics like poly alpha olefins.

Another interesting aspect of turbine oils is the application of the concept of residual life applied successfully by industry. ASTM D4378-84a governs this practice. The RBOT test that determines oxidation life based on pressure drop in an oxygen bomb is used for this purpose. The procedure is as per ASTM D2272. The life is measured in minutes rather than hours and hence is quick to carry out. The field samples are taken at fixed intervals, say, every 5000 hours and then subjected to this test. The life is compared to that of the fresh oil and oil rejection is fixed at 30 to 40% of the original oil. At any stage, the remaining life can also be assessed assuming an exponential change with time. The methodology followed taking into account the topping rate has been described [25]. Exponential decay implies pseudo first order kinetics, which is a reasonable assumption. The test procedure is important from the methodology point of view. This is an example in which a real system is evaluated periodically by means of a laboratory technique and correlated with practice. Besides oxidation, air release, foaming, and other tests are also done and should be within acceptable limits as laid down. The oxidation test is selected because of its predictive capability of the real system.

Steam and gas turbine oils are undergoing changes and more effective additives like amines are being used increasingly with or without DBPC. Base stocks are improving significantly and these systems, along with rust inhibitors and other additives may have to be reexamined for interactions, if any. Also the existing methods to evaluate oxidation stability may or may not be adequate. Thus the evaluation methods need continuous updating. The difficulties involved in fundamental modelling make this inevitable at least in the near future.

2.5.2 Evaluation of engine oils

Simulation has been successful in one area of engine oils. This relates to the performance aspects related to viscosity. Cold cranking as well as pumping behaviour can be well assessed in the laboratory using rotary viscometers. Such simulations involved development over a long period of time. Similarly, the high temperature, high shear, viscosity specifications are based on the realisation that viscosity requirements in bearings are governed by effective viscosity at the relevant shear rates rather than kinematic viscosity. If the viscosity changes during operation are not large, the information obtained with fresh lubricants is adequate for prediction. Deeper understanding of rheology is also of relevance to the energy losses and it is now possible to at least predict viscous losses from fundamental considerations [26]. Boundary friction losses are more difficult to predict though some attempts are being made with laboratory friction machines. Simulations are successful in this area because the physical property involved, and its variations, can be predicted on a fundamental basis.

Performance characteristics that depend on chemical changes like sludge, carbon deposits, and wear are now being assessed through engine tests due to the inability to simulate this complex situation in the laboratory. Different additive-additive interactions have been summarized by Spikes [27]. When the interaction with the base oils and the environment are also taken into account, the mechanisms become far more involved. Major issues related to the state of art in engine oils and other automotive lubricants are well covered in a book edited by Bartz [28]. Attempts are continuing to develop better laboratory techniques for simulation. One recent example is the evaluation by D4742 for gasoline engine oils. The oxidation is conducted under oxygen pressure in a bomb in the presence of water, catalyst, and oxidised fuel. The induction time related to a sharp pressure drop forms the test criterion. It is of interest to observe that 'partial simulation' is attempted with fuel preoxidised in the presence of NO_2 as these components come in contact with the oil in engines via blow-by. The method states that the technique is able to correlate with viscosity increase in engine sequence tests but cautions that the test is not a substitute for engine tests. Such tests may be able to screen formulations to a limited extent. Laboratory test methods of various kinds have been developed over several decades and will not be gone into here. Most of the tests are not successful and have limited possibilities.

It is no exaggeration to state that the effective simulation for engine oils is the engine itself. With this real situation, can there be any other approach to screening? One possibility is the development of laboratory tests centred around the engine tests. During engine tests, samples can be withdrawn at different time intervals and tested in a known or developed laboratory test. The trend observed should be capable of predicting the final test result in the engine. This is based on the reasonable assumption that precursors to deterioration, as well as deteriorated products, accumulate with time. Thus in an engine test run for 200 hours to rate deposits, a laboratory test with samples drawn say every 2 hours may adequately predict trends in 20 hours. The test methods may have to be different and should not be restricted to 'standard' tests. This approach coupled with detailed chemical analysis, can provide better fundamental understanding since real changes are being monitored. If successful, they provide opportunity to screen formulations with shorter duration engine tests. In a simpler system like turbine oils trends in oxidation are predictable by evaluating field samples periodically as stated earlier. It is considered that this approach is worth considering for complex systems as well. One recent paper by Kollmann et al [29] is of relevance here. In this paper the authors emphasise the fact that oil drain interval is a function of operating conditions and proposed a monitoring system based on electrical conductivity. The monitoring can be done on-line and provides information to the user regarding oil rejection. The methodology is of importance as this is based on the recognition that oil deterioration can be predicted only on the basis of the changes in the engine itself. Several investigations are reported in literature in which used oils at different intervals are analysed. However, the suggestion here is to use samples for trend analysis via laboratory tests.

2.5.3 Tribological implications

Many tribological contacts operate in the mixed and boundary regimes leading to friction and wear in the contacts. Friction can be reduced to some extent by special additives called friction modifiers. The aspect of fluid film friction is not being considered here. While boundary friction control is useful the major problem of boundary contacts is wear which needs to be controlled. The concepts involved will be discussed in chapters 4 and 5. Wear control in lubricated contacts is an integral part of the overall lubricant performance. In many engineering situations, wear control is accomplished by antiwear additives. These additives are effective only if their interaction with the surfaces leads to wear resistant films. In complex systems, surface interactions can occur with other additives, as well as oxidised products, interfering with the antiwear performance. The other problem is that wear testing is in many cases carried out in standard test machines, which simulate neither the contact conditions nor the materials. Thus, more often than not there is no correlation between laboratory test and the real wear in a component. One example is a study of engine oil oxidised to different levels and tested for wear in a cam-and-tappet rig and a 4-Ball wear tester [30]. No correlation was observed between the two. In another example the influence of dithiophosphtes containing different metals and side chains has been evaluated by four different tribological tests [31]. The products were rated for performance in the 4-Ball EP and antiwear testers as well as in an FZG rig for wear and EP properties. The authors found different ratings with different machines. They also found that the ratings with a given machine itself can change depending on the criterion used. These examples show that characterizing wear and the associated problem of load carrying capacity even for fresh lubricants is ill defined. This raises questions with regard to practical use of large amount of work on EP and antiwear additives simply rated by 4-Ball machines. Fundamental studies done on the reaction mechanisms concentrate on composition of chemical films and their thickness. The link between film composition and wear is not yet clear. Understanding of additive-additive and additive-base oil interactions with regard to these additives in complex systems is not yet available. The mechanisms involved in running-in are yet poorly defined. Under these circumstances, it is necessary to look comprehensively into wear and related problems with an aim to improve practice. Besides lubricants many other parameters like materials, surface roughness, and oxide films affect wear. It is hence important to address all these problems and bring theory closer to practice. This is the main objective of the present book.

2.6 Synthetic lubricants

Synthetic lubricants may be defined as those fluids, which are chemically synthesised unlike mineral oils obtained from crude. Synthetic lubricants account for about 2% of the global lubricant market which is growing. They are used in many advanced applications like aviation and space where mineral oils cannot function. Synthetics are also being used in the more traditional applications like engine oils. This section considers the categories of lubricants and their application.

2.6.1 Types of synthetic lubricants

A state of the art consideration of synthetic lubricants has been presented in a recent book [32], which considers both applications and chemistry. Many types of synthetic lubricants are now available and each type has advantages and disadvantages in comparison to a mineral oil. A good comparison of synthetic fluids is given by Bartz [33] with a 1 to 5 scale of decreasing performance. This information is reproduced in Table 2.3. The comparison is directional and provides an approximate rating. It shows that all synthetics have some desirable and undesirable properties and the selection is based on the specific need and the expected properties. It can be seen that perfluoroalkyl ethers are outstanding with regard to oxidation and thermal stability, but deficient with regard to corrosion, compatibility with mineral oils, and biodegradability. In addition, the cost of the fluid can be 500 times that of mineral oil. Less expensive fluids like diesters, and polyalpha olefins (PAO) cost 3-5 times the mineral oil and are increasingly being used for automotive and industrial applications. The structures of commonly used fluids are given in Fig. 2.5.

2.6.2 Applications

Formulated lubricants based on PAO and diesters are finding increasing use in engine oils, particularly in Europe. The formulations may be based fully on the synthetic fluid or partially mixed with the mineral base fluid. PAO, as can be seen from Table 2.3, has very good VI that normally exceeds 130, excellent low temperature properties, and very good thermal and oxidation stability. The requirements for VI, HTHS viscosity, and low temperature properties for engine oils are already considered. PAO is superior in all these properties and can reduce or even eliminate VI improvers. Another aspect is the significantly lower volatility of PAO in comparison to mineral oils of similar viscosity. Volatility is of increasing concern in Europe to reduce pollution and oil consumption. Some engine specifications include the volatility levels, as per the DIN 5158 NOACK method. PAO is thus well suited for this purpose. It also has superior oxidation stability and this contributes to the longer life of the oil. Another advantage is that PAO, being a hydrocarbon is completely miscible with mineral oil and can be partially blended with it. Diesters, termed as dicarboxylic acid esters in the table, are also used and are comparable to PAO. They are deficient in hydrolytic stability and seal compatibility and these properties have to be taken into account during use. Biodegradability of diesters is superior to PAO and mineral oils and hence they are finding increasing use in 2-Stroke engines where oil is premixed with the fuel. PAO and diesters are also making in-roads into industrial lubricants. A recent unique example is the successful replacement of water based cutting fluids by neat diester fluids by Dailmer-Benz [34]. Despite the high initial cost, the overall benefit in terms of pollution control and work quality result in a better cost to benefit ratio. Several known applications and new possibilities exist for various synthetics. These include fire resistant hydraulic fluids, special compressor oils, and gear oils. Polyalkylene glycols, polyisobutylenes and esters are being more widely used in metal cutting and forming operations because of their low staining

Table 2.3 Comparison of synthetic lubricants (From Ref. [33], p335 by courtesy of Marcel Dekker Inc.)

Evaluation 1– Excellent, 2– Very good 3– Good, 4– Moderate 5– Poor	Mineral oils	Polyisobutenes	Polyalphaolefins	Alkylated aromates	Polyalkylene glycols.	Perfluoroalkylethers	Polyphenylethers	Dicarboxilic acid esters	Neopentyl polyesters	Triaryl phosphate esters	Trialkyl phosphate esters	Silicone oils	Silicate esters	Silahydrocarbons	Chlorofluorocarbons	Cyclophosphazene fluids	Dialkylcarbonates	Alkylated cyclopentanes	PAMA/PAO-Cooligomers	Rapseed oils
Viscosity temperature behaviour (VI)	4	5	2	4	2	4	5	2	2	5	1	1	1	2	4	5	3	3	2	2
Low-temperature behaviour (pour point)	5	4	1	3	3	3	5	1	2	4	1	1	2	3	3	3	3	3	2	3
Liquid range	4	5	2	3	3	1	5	2	2	2	3	1	1	2	5	4/5	2	1	2	3
Oxidation stability	4	4	2	4	3	1	2	2/3	2	2	4	2	2	3	1	3	2	2	2	5
Thermal stability	4	4	4	4	3	1	1	3	2	2	3	2	3	2	2	3/4	3	4	3	4
Evaporation losses, volatility	4	4	2	3	3	1	3	1	1	2	2	2	3	2	3	3	4	1	1	3
Fire resistence, inflamability temperature	5	5	5	5	4	1	4	4	4	1/2	1/2	3	4	4	1	1/2	3	5	4/5	5
Hydrolytical stability	1	1	1	1	3	1	1	4	4	4	3	3	4	1	2	3	3	1	2	5
Corrosion protection properties	1	1	1	1	3	5	4	4	4	4	4	3	5	1	5	3	1	1	2	1
Seal material compatibility	3	3	2	3	3	1	3	4	4	5	5	3	3	2	4	3/4	3	2	1	4
Paint and lacquer compatibilty	1	1	1	1	4	2	4	4	4	5	5	3	4	1	3	3/4	2	1	1	4
Miscibility with mineral oil	—	1	1	1	5	5	3	2	2	4	4	5	4	1	5	5	2	1	1	1
Solubility of additives	1	1	2	1	4	5	2	2	2	1	1	5	3	3	5	4	2	3	1	3
Lubricated properties, load-carrying capaciy	3	3	3	3	2	1	1	2	2	1	3	5	4	3	1	2/3	2	3	2	1
Toxicity	3	1	1	5	3	1	3	3	3	4/5	4/5	1	4	2	2	2	1	1	1	1
Biodegradability	4	5	5	5	1/2	5	5	1/2	1/2	2	2	5	4	5	5	—	1	5	4/5	1
Price compared with mineral oil	—	3-5	3-5	3-5	6-10	500	200-500	4-10	4-10	5-10	5-10	30-100	20-30	30-70	300-400	30-50	4-10	3-8	5-10	2-3

Class	Typical structural formula
Synthetic hydrocarbon (Polybutene)	$(-CH_2-CH_2-CH_2-CH_2-)_n$
Chlorofluorocarbon	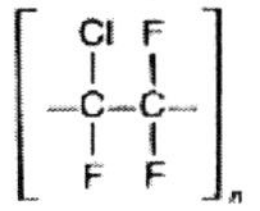 $\left[-\overset{Cl}{\underset{F}{C}}-\overset{F}{\underset{F}{C}}-\right]_n$
Diester	$C_8H_{17}-O-CO-C_8H_{16}-CO-O-C_8H_{17}$
Neopentyl polyol ester	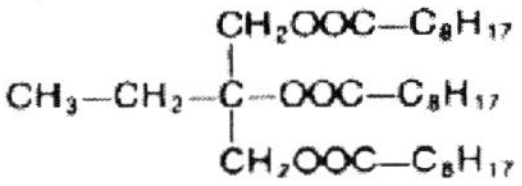 $CH_3-CH_2-\overset{CH_2OOC-C_8H_{17}}{\underset{CH_2OOC-C_8H_{17}}{C}}-OOC-C_8H_{17}$
Fatty acid ester	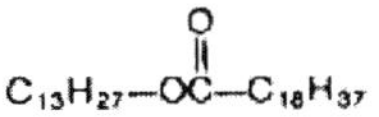 $C_{13}H_{27}-O\overset{O}{\overset{\|}{C}}-C_{18}H_{37}$
Polyglycol ether	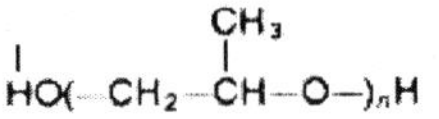 $HO(-CH_2-\overset{CH_3}{CH}-O-)_nH$
Fluoroester	$F(CF_2)_4CH_2OOC(CF_2)_4F$
Phosphate ester	$(CH_3-C_6H_4-O)_3P=O$
Silicate ester	$Si(O-C_8H_{17})_4$
Disiloxane	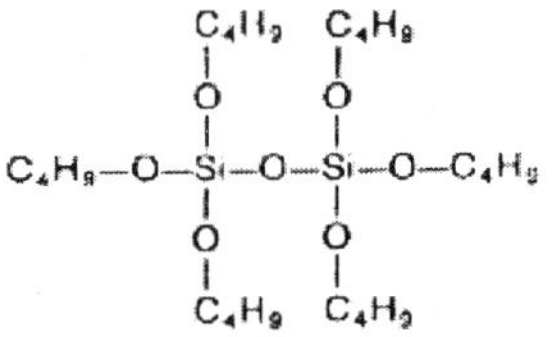 $C_4H_9-O-\overset{C_4H_9-O}{\underset{O-C_4H_9}{Si}}-O-\overset{C_4H_9-O}{\underset{O-C_4H_9}{Si}}-O-C_4H_9$
Silicone	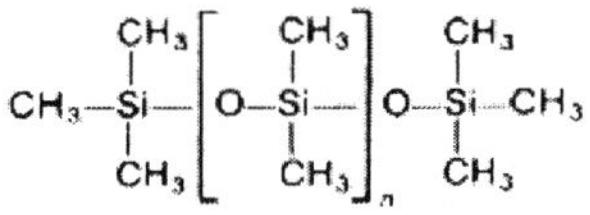 $CH_3-\overset{CH_3}{\underset{CH_3}{Si}}-\left[O-\overset{CH_3}{\underset{CH_3}{Si}}-\right]_n O-\overset{CH_3}{\underset{CH_3}{Si}}-CH_3$
Silane	$(C_{12}H_{25})Si(C_6H_{13})_3$
Polyphenyl ether	 $C_6H_5-O-C_6H_4-O-C_6H_5$
Perfluoroalkyl polyether	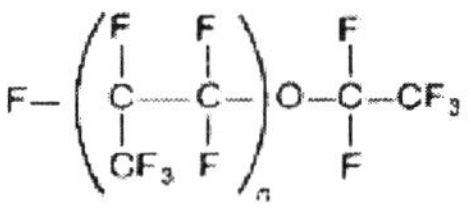 $F-\left(\overset{F}{\underset{CF_3}{C}}-\overset{F}{\underset{F}{C}}-\right)_n O-\overset{F}{\underset{F}{C}}-CF_3$

Fig. 2.5. Typical structures of major synthetic fluids.
(From Ref. [18] p169)

and clean burn-off characteristics. Another important application is for aviation gas turbines. Polyol esters are the main lubricants meeting the stringent DEngRD 2497 specification. These lubricants are manufactured only by a few companies in the world. The specification is rigorous with regard to oxidation stability, deposits and performance in gear rigs. Other fluids mentioned in the table are less used due to cost as well as application problems. Silicones are very stable but have poor lubrication properties. Silicone based greases are used to a limited extent in space and related applications. Silahydrocarbons are a relatively new class in which carbon is replaced by silicon and attached to alkyl groups. Polyphenyl ethers, as well as perfluoroalkyl ethers have potential application for very high temperatures exceeding 300°C. One recent application of perfluoro ether is its use in the computer disc-head interface to prevent stiction [35]. The film applied is only 2-3 nanometers thick and is effective.

Tribological implications are directionally similar to those raised with regard to mineral based lubricants. Synthetics that are polar generally have better tribological properties. When additives are added, their compatibility with the base fluid is important and in each case, this has to be ensured. One extreme example is silicone fluid whose tribological properties cannot be improved with known additives. Research is being conducted to develop special additives for silicone fluids.

2.7 Environmental issues

The lubricants used are disposed of in various ways. A survey made of the European scene [36] shows that out of 2.6 million tons of reusable used lubricants, 1.1 million tons is unaccounted for, and may have been indiscriminately disposed of finding its way into the environment. Similar problems exist in the US and other countries. There are also situations in which loss into the environment is unavoidable as in mobile out-board equipment, metalworking, and other operations. The lubricants, with different degrees of deterioration, eventually end up as pollutants in the ecosystem. The hazards involved have to be assessed. Such pollution can be particularly harmful to aquatic life. The ecological problems are very involved and many answers are partial though in the desirable direction. Methods to evaluate hazardous chemicals are available. These methods are adapted for lubricant evaluations with modifications. Three aspects to be considered are toxicity, bioaccumulation, and biodegradability. Toxicity is evaluated on the basis of lethal effect on selected organisms at a particular dosage within a prescribed time period. The product is considered lethal if 50% of the population is killed in the specified time period. The German DIN specifications are well known in this area. These specifications describe procedures for mammalian, fish, and bacterial toxicity. Fish toxicity is of particular importance to aquatic life and is determined by DIN 38 412-15. Water hazard or WGK is based on the water endangering number (WEN) that is referred to as WKZ in German specifications. The procedure depends on finding an average WEN based on the individual WEN values for the three toxicities referred to above. For fish and bacterial toxicity it is found as follows:

$$WEN = (-\log NEC \text{ in ppm}/10^6 \text{ ppm})$$

where NEC refers to 'no effect concentration'. The WEN for mammalian toxicity on the other hand is obtained by assigning a number based on lethal toxicity. The hazard levels are then classified on the basis of the ranges of WEN values.

It may be noted here that these specifications are constantly evolving and the above discussion only presents the central approach to quantify toxicity.

Bioaccumulation is the concentration of the product from water or food. This is related to the metabolism and is assessed by bioconcentration factor. Biodegradability refers to the ability of naturally occuring microorganisms to degrade a product. This is the main aspect being studied with regard to lubricants. Biodegradability is assessed by the CEC-L-33-A-93 test procedure. In this test a defined amount of fluid is added to mineral media inoculated with microorganisms from natural sources. The mixture is shaken in the dark for 21 days and degradation measured by the loss of infrared bands. ASTM D5864 is based on a similar approach though the criterion used to measure degradation is based on the evolution of CO_2. A value of 70% degradation in CEC test rates the material as readily biodegradable.

Mineral oils are not considered toxic. But some chemical additives like zinc dithiophosphtes and calcium alkaryl sulphonates are considered toxic and substitutes for these additives are being developed. The issue of bioconcentration has not been well studied with regard to lubricants. Biodegradability of mineral oils is not high and many fluids fall below 70% limit. This has led to the development of easily degradable vegetable oil based lubricants for some applications. Some synthetic fluids like diesters also are easily biodegradable. The tribological performance of these lubricants is also being studied [37,38]. Many countries are encouraging ecofriendly lubricants through ecomarks like 'Blue Angel' in Germany. Such marking demands very low toxicity and biodegradability of greater than 80%. The market for eco-friendly lubricants is growing and several formulations based on rapeseed oil, ester, and polyglycols are now available. The major interest in these lubricants is at present in Germany, Austria, and Switzerland. There is a significant effort in various countries with regard to testing environmental effects and it is likely that global standards may emerge in the near future. The discussion of environmental issues here has been limited to lubricants and the issues of air pollution from engine exhaust emissions is not considered. Lubricants also contribute to exhaust emissions due to evaporation. There is also evidence that lubricants contribute to particulate emissions in diesel engines [39]. Another aspect of importance is the carcinogenicity of used engine oils. It is known that polynuclear aromatics produced in the combustion process accumulate in the engine oils [40]. These products are known carcinogens and can pose disposal problems.

Centralised collection and disposal in an organised manner is encouraged in many countries through legislation and awareness programmes. But still a large amount of oil is disposed off indiscriminately. The oils collected in the organised sector are utilised for heating and other purposes as well as re-refining. Nearly 20-30% of used engine oils are re-refined and the base stocks obtained are blended with additives and sold. Such re-refining is very desirable as the lubricant is effectively recycled. The re-refining technology poses problems. The acid and clay treating process, which is easy to implement for small-scale operations, has environmental problems due to acid sludge and used clay disposal. In addition, modern lubricants are drained after long usage and need stronger treatments resulting in lower yield of base oil. Processes, which use solvent extraction and/or hydrotreatment, are economical only for large-scale plants. It is hence not surprising that the acid clay route is still an important process for re-refining in many countries. With ecological concerns building up there is incentive to develop cost effective technology for small-scale operations. In the past, several re-refining methods have been tried and some of the ideas involved may be worth reconsidering. One can also speculate on the possibility of the used oils being pre-treated and directly used in lube refineries in deasphalting or other units. Used oil can form a small percentage of the existing feed. Such possibilities need detailed assessment.

Biodegradable lubricants do have their limitations. Synthetic fluids are expensive as discussed in the previous section. Oxidation stability of fatty oils is poor and they can be used only at moderate temperature. In some countries, use of vegetable oil is mandatory for some applications like chain saw lubrication. But large number of industrial applications will continue to be based on mineral oils and the parallel developments in collection and re-use technologies is essential.

References

1. R. N. Watkins, Petroleum Refinery Distillation, Gulf Publishing, Houston, 1973.
2. J. D. Seader, J. J. Siirola, and S. D. Barinicki, Petroleum and complex mixture distillation, in Perry's Chemical Engineers' Handbook, 7th edition, McGraw-Hill, New York, 1997, 13-85.
3. R. T. Aird and S. L. Forgham, The lubricating quality of aviation fuels, Wear, 18 (1971) 361.
4. R. A. Vere, Lubricity of aviation turbine fuel, SAE Paper No. 690667 (1969) 2237.
5. D. Cooper, Laboratory screening tests for low sulphur diesel fuel lubricity, Lub. Sci., 7 (1995) 133.
6. S. Ikejima, H. Sugi, T. Ichikawa, K. Taniguchi, and K. Saito, Study of diesel fuel lubricity, JSAE Review, 18 (1997) 203.
7. A. Dorinson and K. C. Ludema, Mechanics and Chemistry in Lubrication, Tribology series 9, Elsevier, 1985, Chapter 16.
8. American Petroleum Institute Technical Data Book- Petroleum Refining, API, Washington, 1983.
9. M. R. Gray, Upgrading Petroleum Residues and Heavy Oils, Marcel Dekker, New York, 1994.
10. A. G. Bridge, Hydrogen processing, in R. A. Meyers (ed.), Handbook of Modern Petroleum Technology, 2nd edition, McGraw-Hill, New York, 1997, Chapter 14.1.
11. A. Sequeira. Jr, Lubricant Base Oil and Wax processing, Marcel Dekker, New York, 1994.
12. F. Buyan, Mobil hydrodesulphurisation process for distillates, Proc. Mobil Technology Seminar, Tata McGraw-Hill, New Delhi, 1995, 173.
13. Methods of Test for Petroleum Products and Lubricants, 5.01 to 5.04, American Society for testing Materials, 1998.
14. C. V. Smalher and R. K. Smith, Lubricant Additives, The Leizus-Hiles Co,Cleveland, 1967.
15. M. W. Ranney, Lubricant Additives- Recent Developments, Noyes Data Corporation, New Jersey, 1978.
16. E. R. Booser (ed.), CRC Handbook of Lubrication, Vols. I and II, CRC Press, 1983.
17. R. M. Mortier and S. T. Orszulk (eds.), Chemistry and Technology of Lubricants, Blackie, Glasgow, 1992.
18. E. R. Braithwaite (ed.), Lubrication and Lubricants, Elsevier, Amsterdam, 1967.
19. E. S. Nachtman and S. Kalipakjian (eds.), Lubricants and Lubrication in Metalworking Operations, Marcel Dekker, New York, 1985.
20. J. P. Byers (ed.), Metal Working Fluids, Marcel Dekker, New York, 1994.

21. B. Basu, D. Saxena, V. Kaul, M.I. S. Sastry, and R. T. Mokken, Prediction of oxidation stability of inhibited base oils using an artificial neural network (ANN), Lub. Sci., 10 (1998) 121.
22. D. C. Kramer, B. K. Lok, and R. R. Krug, The evolution of base oil technology, in W. R. Herguth and T. H. Warne (eds.), Turbine Lubrication in the 21st Century, ASTM STP 1407, ASTM, 2001.
23. J. Igarashi and T. Yoshida, Computer simulation of turbine oil oxidation 1: consumption of a hindered phenol antioxidant in model hydrocarbon systems at 115°C, Lub. Sci., 7 (1994) 3.
24. J. Igarashi and T. Yoshida, Computer simulation of turbine oil oxidation 2: consumption of a hindered phenol antioxidant at 115^0C, Lub. Sci. 7 (1995) 107.
25. M. J. Denherder and P. C. Vienna, Control of turbine oil degradation during use, Lub. Eng., (1981) 67.
26. C. Bovington, V. Anghel, and, H. A. Spikes, Predicting sequence VI and sequence VI A fuel economy from laboratory bench tests, SAE Paper No. 972925 (1997).
27. H. A. Spikes, Additive- additive and additive – surface interactions in lubrication, Lub. Sci., 2 (1989) 3.
28. W. J. Bartz (ed.), Engine Oils and Automotive Lubrication, Expert Verlag GmbH, Berlin, 1993.
29. K. Kollman, T. Gürtier, K. Land, W. Warnecke, and H. D. Müeller, Extended oil drain intervals- conservation of resources or reduction of engine life (Part II), SAE Paper No. 981443 (1998) 728.
30. G. Monteil, A. M. Merillon, J. Lonchampt, and C. Roques-Carmes, Evaluation of the antiwear performance of the aged oils through tribological and physicochemical tests, Lub. Sci., 4 (1992) 155.
31. M. Born, J. C. Hipeaux, P. Marchand, and G. Parc, The relationship between chemical structure and effectiveness of some metallic dialkyl and diaryl dithiophosphates in different lubricated mechanisms, Lub. Sci., 4 (1992) 93.
32. R. L. Shubkin (ed.), Synthetic Lubricants and High-Performance Functional Fluids, Marcel Dekker, New York, 1993.
33. W. Bartz, Comparison of synthetic fluids, in R. L. Shubkin (ed.), Synthetic Lubricants and High-Performance Functional Fluids, Marcel Dekker, New York, 1993, Chapter 14.
34. Theo Mang, Future trends in machine tool lubrication, in D. Basu and S. P. Srivastava (eds.), Proc. Intl. Sym. on Fuels and Lubricants, Tata McGraw-Hill, New Delhi, 1997, 49.
35. B. Bhushan and Z. Zhao, Macroscale and microscale tribological studies of molecularly thick boundary layers of perfluoropolyether lubricants for magnetic thin film rigid disks, J. Info. Storage Proc. Syst., 1(1999) 1.
36. CONCAWE, Report No. 5, Collection and disposal of used lubricating oil, 1996.

37. B.- R. Höhn, K. Michaelis, and R. D. Döbereiner , Load carrying capacity properties of fast biodegradable gear lubricants, Lub. Eng., 55 (1999) 15.
38. W. Belluco and L. De Chiffre, Testing of vegetable-based cutting fluids by hole making operations, Lub. Eng., 57 (2001) 12.
39. P. Tritthart, F. Ruhri, and W. Cartellieri, The contribution of the lube oil to particulate emission of heavy duty diesel engines, in W. J. Bartz (ed.), Engine Oils and Automotive Lubrication, Expert Verlag GmbH, Berlin, 1993, Section 4.8.
40. P. Van Donkelaar, Environmental effects of crankcase and mixed lubrication, in W. J. Bartz (ed.), Engine Oils and Automotive Lubrication, Expert Verlag GmbH, Berlin, 1993, Section 4.11.

3. Dry wear mechanisms and modelling

3.1 Introduction

Wear occurs in tribological components due to several processes that result in the loss of material. The progressive loss of material leads to dimensional changes and eventual loss of performance. Replacement of worn components involves the cost of the new components as well as expenses involved in maintenance and down time. There is an obvious economic advantage in reducing wear and prolonging the life of components. Understanding of wear mechanisms and the ability to predict wear by appropriate models can lead to better control of wear. With this aim in the background a large amount of fundamental and applied research is being conducted in this area. The emphasis has been on dry wear of materials though lubricated contacts are industrially far more common. Several specialized textbooks are available dealing exclusively with wear of materials some of which may be cited [1,2,3].

The generation and removal of wear particles is a complex process. The mechanisms involved are conventionally classified into adhesive, abrasive, fatigue, and oxidative wear. In many engineering situations two or more mechanisms may be operative and it is difficult to model such situations. The separation into categories is essential for understanding individual mechanisms. Controlled laboratory studies can be conducted where one mechanism dominates leading to an understanding of that mechanism. Here again the presumed mechanism cannot be assured leading to uncertainty in the understanding. Despite this scenario a broad understanding of mechanisms is possible. The purpose of this chapter is to provide the necessary background on dry wear of materials. Dry wear in the present context refers to situations where no external lubricant is applied. The coverage is limited to an appreciation of the main concepts involved. Some of the mechanisms will be elaborated in later chapters.

The first two sections deal with contact stresses and asperity temperatures that are of importance both for dry and lubricated wear. Adhesive wear phenomena are then considered followed by a consideration of the available models. Abrasive wear is then covered in a similar manner. As fatigue wear will be covered later in a separate chapter the coverage is limited to a few general observations. This is followed by a consideration of oxidative wear. The final section deals with wear maps. All these sections will deal only with metallic materials. Dry and lubricated wear of non-metallic materials is covered in a separate chapter.

3.2 Contact stresses

Contact stresses and their distribution is of importance in tribology. This section provides basic formulae for stress distributions in concentrated point and line contacts. Concentrated contacts are also referred to as Hertzian contacts. Detailed consideration of contact mechanics is not within the scope of the book and the interested reader can consult specialised textbooks [4,5] in the area.

3.2.1 Surface Stresses

Concentrated contacts that are non-conformal are characterised by high stresses. Point contact and line contact are the basic contacts considered here. The results presented here are for elastic contacts and smooth surfaces. The point and line contacts are illustrated in Fig. 3.1.

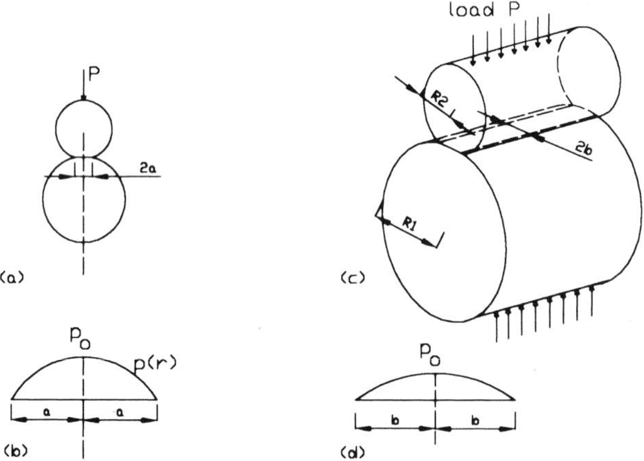

Fig. 3.1. (a) Spheres in elastic contact and (b) semi-elliptical pressure distribution; (c) cylinders in parallel contact and (d) semi-elliptical pressure distribution.

Under load the surfaces deform elastically with typical pressure distributions as shown in this diagram. The loading in both cases is static. For two spheres of radii R_1 and R_2 in contact under a load P

$$\text{contact radius, } a = \left[\frac{3PR}{4E}\right]^{1/3} \tag{3.1}$$

where ν_1 and ν_2 are the Poisson ratios of the two materials and E_1 and E_2 are the elastic modulus of the two materials.

Also, $\dfrac{1}{E} = \dfrac{1-\nu_1^2}{E_1} + \dfrac{1-\nu_2^2}{E_2}$ and $\dfrac{1}{R} = \dfrac{1}{R_1} + \dfrac{1}{R_2}$

The elliptical pressure distribution is

$$p(r) = p_0\left[1 - \frac{r^2}{a^2}\right]^{1/2} \tag{3.2}$$

where r is any radius within the contact circle

The maximum pressure p_0 occurs on the axis of symmetry and is

$$p_0 = \frac{2P}{\pi a}$$

The mean pressure $p_m = \dfrac{2}{3} p_0$

The contact deformation $\delta = \dfrac{a^2}{R}$

With these formulae the equations (1.9a) and (1.9b) in chapter 1 for asperity deformation can be easily derived.

For two parallel cylinders in line contact with load P now expressed per unit length

$$\text{half contact width } b = \left[\frac{2PR}{\pi E}\right]^{1/2} \tag{3.3}$$

$$\text{pressure distribution } p(x) = p_0\left[1-\frac{r^2}{b^2}\right]^{1/2} \tag{3.4}$$

$$\text{also peak pressure } p_0 = \frac{3P}{2\pi a^2}$$

$$\text{and mean pressure } p_m = \frac{\pi p_0}{4}$$

$$\text{central deformation } \delta = \frac{\left(1-v_1^2\right)\left[l_n\left(\frac{4R_1}{b}\right)-\frac{1}{2}\right]}{E_1} + \frac{\left(1-v_2^2\right)\left[l_n\left(\frac{4R_2}{b}\right)-\frac{1}{2}\right]}{E_2} \tag{3.5}$$

Formulae are available in standard texts to calculate stress distributions with arbitrary contact geometries and are not given here. Stress distributions at any point in the sub-surface can also be obtained by the available methods [4,5]. With real surfaces the contacts are at the asperities and this will induce changes in stress distribution as compared to smooth surfaces [6,7]. The overall stress distribution at the surface is not significantly affected for roughness levels commonly used in practice though stresses at contact spots will be higher. Sub-surface stresses in the zone close to the surface are more strongly influenced by local asperity stresses.

3.2.2 Sub-surface stresses

The point of maximum shear stress and its magnitude are of particular importance. For point contact $\tau_{max} = 0.31\,p_0$ and occurs at a depth of 0.48a for steel surfaces on the axis of symmetry. In the case of line contact $\tau_{max} = 0.30\,p_0$ and occurs at a depth of 0.78b for steel surfaces on the axis of symmetry. These values refer to situations where only normal stresses are acting. In many contacts the maximum Hertzian stress is higher than the yield stress and the plastic deformation is initiated at the point of maximum shear stress. This stress is of importance with regard to the initiation of sub-surface cracks.

3.2.3 Frictional traction

When frictional (tangential) force is applied the stress distributions in the contact change. The stresses at any point are obtained by superposing the stresses due to normal and tangential forces provided the Poisson ratios of the two surfaces do not differ significantly. This is normally the case for metallic contacts. Tangential stresses influence the maximum shear stress and its location. The point of maximum shear stress moves towards the surface as frictional force increases and can reach the surface when the friction coefficient f exceeds 0.3. Another important influence of tangential force is the generation of large tensile stresses on the surface. In the case of line contact the maximum tensile stress occurs at the trailing edge of the contact. The trailing edge is with reference to the direction of tangential force. The value is equal to $2f\,p_0$ and is hence substantial. Such stresses can induce surface cracks in the material and contribute to fatigue wear. For point contacts the maximum tensile stress occurs over the trailing edge of the contact periphery. This stress is given by $k^{/}p_0$ where

$$k^{/} = \frac{\pi f(4+v)}{8} + \frac{(1-2v)}{3} \qquad (3.6)$$

where v is Poisson ratio

Here again the tensile stresses can be substantial.

The above treatment is limited and several aspects of contact mechanics are left out. It is hoped this limited treatment provides an adequate background to appreciate stress related problems in tribology. Some aspects will be elaborated upon where necessary in the text. The SI units are commonly used in calculations.

3.3 Asperity temperatures

The problem of surface temperature rise has been considered in section 1.4. The temperature rise was treated on the basis of heat distributed over the geometric contact area. This temperature rise is now designated as $\Delta\theta_g$. Generation of heat is actually at the asperities. It is of interest to estimate the asperity temperature rise $\Delta\theta_a$. The asperity temperature rise can be significantly higher than $\Delta\theta_g$. The high localised temperature rise at asperities affects chemical reactions, material transformations, and several related phenomena. The average asperity temperature rise is high because the heat flux generated over the real area is much higher in comparison to that based on the geometric area. Several approaches with different levels of sophistication are available to deal with the problem. The commonly used simpler approach is adopted here to calculate these temperatures. The first step is to treat the temperature rise at the asperities neglecting the rise of temperature over the geometric area. This is defined as a case with no thermal interaction. The next step is to deal with the situation of thermal interaction leading to more realistic estimation of temperatures. These two cases are considered below.

3.3.1 Asperity temperatures without thermal interaction

The model for contact considered here is based on plastic flow at the asperities. The real contacts can be entirely elastic in which case the areas and heat distribution get modified. The real area will have contact spots of varying diameters and it is usual to assume an average asperity contact diameter. Firstly since real area is known on the basis of load and hardness heat flux is calculated for the real area. This heat flux is flowing through the asperities. Since the average diameter of the asperity is known (or assumed) the Peclet number for the asperity can be calculated. Then the same equations that govern the temperature rise as discussed in chapter 1 are used to estimate asperity temperature rise. Since the thermal interaction is neglected the asperity temperature is obtained by adding the bulk temperature to the estimated temperature. The estimation is illustrated by an example. It may be noted that surfaces I and II in section 1.4 will now be referred to as 1 and 2 wherever applicable. For convenience the case of steel on steel for which overall temperature rise was calculated in section 1.4 is reconsidered here. The contact is a square contact with a geometric area of 1.81×10^{-5} m^2 with half contact width equal to $2.127 \times 10^{-3}\,\mathrm{m}$. The load is 100 N and the sliding speed is 5.11 m/s. For consistency with the previous example calculation is done with a square asperity. In the present example the contact occurs at real area spots distributed within the geometric square area. The parameters related to the asperities carry the subscript 'a' in the following text.

Assume hardness of $3.26 \times 10^{9}\left(\mathrm{N/m^2}\right)$

For the load of 100 N the real area $A_r = \dfrac{\mathrm{Load}}{\mathrm{Hardness}} = 3.067 \times 10^{-8}\left(\mathrm{m^2}\right)$

The calculated heat flux on the basis of real area, $q_a = 6.6 \times 10^{9}\left(\mathrm{W/m^2}\right)$

This value may be compared with q value of 1.12×10^{7} based on the geometric area.

Assume an average square asperity area of 100 $\mu\mathrm{m}^2$. The half contact dimension l_a for the asperity is 5×10^{-6} m. The Peclet number P_{ea} for asperity contact is now calculated to be 1.09. This may be compared with the large Peclet number of 463.7 on the basis of the geometric contact dimension. This is because the asperity dimension is much smaller.

Since q_a and P_{ea} are known the temperature rise at the asperity may be obtained as was done for geometric contact area. In this case the graphical procedure based on Fig. 1.5 is used since P_{ea} is between 0.5 and 5.0. The value obtained for the asperity temperature rise $\Delta\theta_a$ is 256°C. Heat is generated at all the asperities in contact and the temperature rise is same for all of them provided they all have the same dimensions. This temperature rise is much higher than the overall surface temperature rise calculated earlier as 18.1°C. This calculation shows that high asperity temperatures can easily arise in the contact. It may also be noted that for a given heat flux q_a temperature rise increases with the asperity diameter. For example if the average asperity l_a is 20 μm instead of 10 μm the asperity temperature rise is calculated as 436°C. In the process of running-in where some misalignment is possible at the micro level, some asperity contacts may be particularly large. Such asperities may be critical as failure at these contacts can trigger a run away situation that can lead to problems like scuffing. Such problems may be analysed better in terms of critical asperity events rather than 'average' behaviour in the contact. While this simple analysis neglects thermal interaction the temperature may be considered as a lower bound temperature.

In some cases the overall heat generation may be small enough to justify using this approach. For example if the above contact is lubricated the temperature rise at asperities will be much smaller. If *f* in this boundary lubricated contact is 0.1, the temperature rise for a 10 μm asperity will be around 64°C and the overall temperature rise will also be low. For practical purposes the asperity temperature in this case may be obtained by adding the bulk temperature to $\Delta\theta_a$. The asperity contacts are circular and an approximate estimation may be made substituting contact radius for l_a. Tian and Kennedy [8] proposed useful equations for circular contacts both for maximum and average temperatures. The advantage of these equations is that they are generalised for any Peclet number and have a sound theoretical basis. Their general equation for average temperature rise at one surface, $\Delta\theta$ is

$$\Delta\theta_{1,2} = 2qa \text{ x } \frac{0.61}{k_{1,2}\pi^{0.5}\left(0.6575 + P_{e1,2}\right)^{0.5}} \tag{3.7}$$

where q is the heat flux and 'a' is contact radius and applicable for any Peclet number P_e. Subscripts 1, 2 refer to one surface or the other. Adapting this equation for the present case of asperity contact and utilising Eq. (1.17)

$$\Delta\theta = 2q_a a_a \left[\frac{0.61}{k\pi^{0.5}\left\{(0.6575 + P_{ea})^{0.5} + 0.6575^{0.5}\right\}} \right] \tag{3.8}$$

It may be noted $k_1 = k_2$ in this example and is replaced by the general term k. Calculation on this basis gives an average temperature rise of 254°C and is very close to the value of 256°C estimated earlier for a square contact.

The above equation is for a plastic contact area at the asperity. Asperity contacts can be elastic in many cases. For elastic contact the pressure distribution is elliptical and not uniform as in the case of plastic contact. The approach in this case is to obtain q_a on the basis of mean asperity pressure and not the hardness. For example if mean asperity pressure p_m is 0.4H then q_a will be 2.64×10^9. The general equation for elastic contact for the present case from the same reference [8] is

$$\Delta\theta_{1,2} = 2qa \left[\frac{0.732}{k_{1,2}\pi^{0.5}(0.874 + P_{e1,2})^{0.5}} \right] \tag{3.9}$$

Adapting this equation for the present case

$$\Delta\theta = 2q_a a_a \left[\frac{0.732}{k\pi^{0.5}\left\{(0.874 + P_{ea})^{0.5} + 0.874^{0.5}\right\}} \right] \tag{3.10}$$

The value calculated on this basis is 108°C and is substantially lower than 256°C for the plastic case. Well-run engineering surfaces will have mainly elastic contacts and this is advantageous from temperature point of view. For elastic contact the number of contacts and their area can be estimated from the contact theory given in chapter 1. From this information average size and mean pressure at the asperities can be obtained.

It is difficult to decide whether average rise or maximum rise governs the tribological behaviour. It is also difficult to understand the relative importance of overall temperature rise and asperity temperature rise. The temperature rise at the contact is also referred to as flash temperature.

3.3.2 Thermal interaction

While the rise in asperity temperature can be estimated by the above procedure it is necessary to find the actual asperity temperature. If the thermal interaction is neglected this temperature is simply obtained by adding the bulk temperature to the asperity rise. In many lubricated contacts with low heat generation this procedure is adequate. In the case of dry contacts with high heat flux and no cooling by the fluid, thermal interaction becomes more important. Thermal interaction refers to the overall rise in the surface temperature due to the contribution from all the asperities distributed over the geometric contact area. This rise has to be superimposed additionally to obtain the asperity temperature. The asperity temperature θ_a is then

$$\theta_a = \Delta\theta_a + \Delta\theta_g + \theta_b \tag{3.11}$$

where

$\Delta\theta_a$ = temperature rise due to asperity interaction alone
$\Delta\theta_g$ = temperature rise over the geometric area
θ_b = bulk temperature assumed here to be the same for both the bodies

The main approach is to consider that the asperity temperature in the contact is the same for the contacting asperities. The heat is partitioned between the two surfaces in such a way that the following equality results

$$(\Delta\theta_a + \Delta\theta_g)_{\text{body 1}} = (\Delta\theta_a + \Delta\theta_g)_{\text{body2}} \tag{3.12}$$

The asperity temperature rise that now includes thermal interaction is designated as $\Delta\theta_{at}$ and is equal to the sum expressed above. In this particular case the heat flowing to the two bodies will be based on the partition of the heat flux q into the two bodies. The problem is to find $\Delta\theta_{at}$ which will be referred to as asperity temperature rise with thermal interaction.

The approximate method is based on the previous approach. First it is considered that all heat is flowing into body 1. The temperature rise at the asperity and geometric area at the respective Peclet numbers is calculated and summed. This gives $\Delta\theta_{at1}$. Similarly $\Delta\theta_{at2}$ is obtained for the surface 2. The average temperature rise $\Delta\theta_{at}$ is then obtained using Eq. 1.17. In the following equations it is convenient to express temperature rise for asperity as well as well as geometric area in terms of heat flux based on real area. This is done by defining two non-dimensional parameters as

$$l^* = \frac{l_a}{l_g} = \text{half width of asperity contact/half width of geometric contact}$$

$$A^* = \frac{A_r}{A_g} = \text{real area/geometric area}$$

This is now applied to the multiple contact situation as in the previous case but with thermal interaction. Consider the whole heat flowing into body 2. The temperature rise at the asperity contact will be

$$\Delta\theta_{at2} = \frac{0.946 q_a l_g l^*}{k_2} + \frac{0.946 q_a l_g A^*}{k_2} \tag{3.13}$$

This equation adds the temperature rise over geometric area to include the interaction effect. The term $q_a A^*$ represents the heat flux over the geometric area.

For body I assuming high speed equations are applicable with Peclet numbers for asperity and geometric area greater than 5.0

$$\Delta\theta_{at1} = \frac{0.75 q_a l_g l^* P_{ea}^{-1/2}}{k_1} + \frac{0.75 q_a l_g A^* P_e^{-1/2}}{k_1} \tag{3.14}$$

where P_e refers to the Peclet number based on geometric area. Peclet numbers are not distinguished by subscripts here as in the present case they are applicable to one surface only.

Similar relation was proposed by Suh [9] for the high speed case with a correction. This correction is to account for the fact that the overall heat generation should exclude the heat generation at the asperity under consideration which is already accounted for by the first term. Usually a large number of asperities are in contact and the effect due to a single asperity is neglected in the above treatment. The asperity temperature rise $\Delta\theta_{at}$ that includes thermal interaction can be obtained from

$$\frac{1}{\Delta\theta_{at}} = \frac{1}{\Delta\theta_{at1}} + \frac{1}{\Delta\theta_{at2}}$$

The high speed equation above is of limited utility. Usually P_{ea} will be much smaller than P_e and so will not be governed by the high speed equation. Also P_e can have any value depending on speed and contact dimension. If it is assumed that the heat flow at the asperity and geometric area are governed by the respective Peclet numbers, the temperature for a moving source may be expressed as

$\Delta\theta_{at1}$ = asperity temperature rise at its P_{ea} + geometric temperature rise at its P_e

The rest of the procedure is the same to estimate $\Delta\theta_{at}$. As an example $\Delta\theta_{at}$ was estimated for square asperity with plastic contact dimension of $l = 5\ \mu m$ for the same conditions as in previous example. The value estimated is 313°C that may be compared with a value of 256°C obtained without thermal interaction. Thus thermal interaction increased the temperature rise by 57°C. The equations can be adapted approximately for circular contacts replacing l by the radius a. From the general equations cited earlier for circular contacts it is possible to adopt them for the temperature evaluation and obtain better estimates. The asperity temperature is obtained by adding the bulk temperature to $\Delta\theta_{at}$. Note that the treatment was confined to a sliding situation in which one surface is stationary and only the Peclet numbers for one moving source were applied. If both surfaces are moving Peclet numbers are to be considered for both the surfaces.

In the present analysis the asperity temperature is equalised. Hence the geometric area temperature rise for the two surfaces will not be the same. With regard to bulk temperature it may be considered as same for both the surfaces for lubricated contacts. Experimental evidence shows that in lubricated contacts the bulk temperature may be nearly equal to the oil bath temperature [10]. The main reason for this is the convective cooling due to the oil that tends to equalise the temperatures. But the temperature rise in the contact itself can be found by the same procedure using the f value for the lubricated case. The dry contact situation is more complex and the bulk temperatures can be different for both surfaces and the problem needs a complete heat transfer analysis [11]. Another factor is the increase in the temperature of the sliding track over the bulk temperature which also is to be taken into account. These details are not considered here. The temperature estimates on the basis of the present analysis are to be considered approximate. Complete thermal analysis is necessary to predict temperatures more accurately.

3.4 Adhesive wear

3.4.1 Phenomena

Adhesive wear occurs when there is mutual transfer of materials in a moving contact. The process involves firstly transfer of the material from one surface to the other at the interacting asperities. The particles can grow by further transfer in repetitive contacts and eventually get removed as loose debris. There can also be back transfer of material from one surface to the other. These phenomena were well demonstrated in the earlier literature [12]. There is also evidence of intermetallic diffusion based on the analysis of wear particles [13]. When transfer occurs on a large scale leading to performance failure of the components the situation is normally referred to as scuffing. Scuffing is due to large scale transfer and its control is of importance in tribological contacts. The term 'adhesive wear' is meaningful only for situations where progressive wear occurs. The transfer and removal processes can reach a quasi equilibrium state. In such cases a steady wear rate can be observed for the interacting surfaces for a given set of operating conditions.

The transfer at the asperity junctions is considered to be due to intermetallic bonding of the metals. The junction so formed can break in the softer material leading to a transferred fragment. Some junctions can break in the harder material as well leading to transfer of the harder material onto the softer material. Transferred particles can also spread due to shearing resulting in patchy films. When the interface of the junction is weaker than either of the metals shearing occurs at the interface resulting in lower adhesive wear. Adhesion in a fundamental sense occurs between any two contacting asperities. Adhesion need not necessarily involve material transfer. Adhesive wear is observed only in those situations where material transfer occurs.

Real surfaces are covered with oxides and any metal-metal bonding can only occur through the oxides. Thus the nature of oxides and their adherence to surfaces strongly influences adhesive wear. The steady wear process may involve removal of transferred material as well as oxides and it is difficult to separate their relative contributions. Thus when we refer to adhesive wear we are assuming it is the rate controlling mechanism. Fig. 3.2 illustrates schematically adhesive wear mechanism between surfaces covered with oxide films. When asperities deform plastically larger adhesion is expected due to increased disruption of the oxide layer.

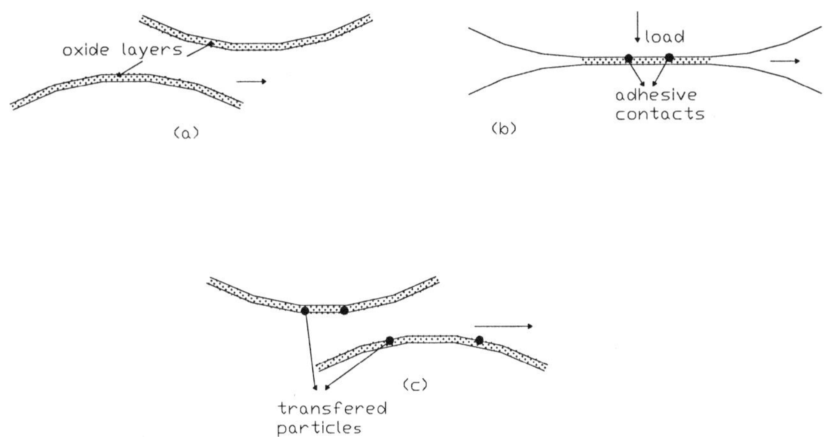

Fig. 3.2. Illustration of adhesive transfer through oxide layers at asperity level: (a) approach (b) contact and (c) disengagement.

3.4.2 Adhesive wear modelling

The complexity of adhesive wear is evident from the previous section. It is difficult to model such a process and to predict wear quantitatively. An early model of adhesive wear is due to Archard [14]. He considered that the wear originated at the real contact spots. Assuming that the real area is plastic, and a hemispherical particle is involved in the wear event, the equation may be derived as follows:

Real area $A_r = \frac{W}{H}$ where W is the load.

Assume average asperity contact diameter d. If a hemispherical wear particle is produced at a contact over a sliding distance d its volume will be

$$v_d = \frac{1}{12}\pi d^3 \tag{3.15a}$$

Since the total number of asperities are equal to real area/individual asperity area it can be shown that the total volume generated at all the asperity contacts over a sliding distance d is

$$V_d = \frac{1}{3}\frac{Wd}{H} \tag{3.15b}$$

If we now define wear rate V_r as wear volume per unit sliding distance it follows

$$V_r = \frac{V_d}{d} = \frac{1}{3}\frac{W}{H} \quad \left(m^3/m\right) \tag{3.15c}$$

The wear volume V is normally known over a sliding distance *l*. In these terms

$$V_r = \frac{\sum V_d}{\sum d} = \frac{V}{l} = \frac{1}{3}\frac{W}{H} \tag{3.15d}$$

The actual wear rates observed are far lower than those obtained from this equation. To obtain realistic values Archard introduced a wear coefficient K and his final equation is

$$V_r = K\frac{W}{H} \tag{3.16}$$

K is a non-dimensional wear coefficient and can vary over a wide range of 10^{-2} to 10^{-5} depending on materials and operating conditions. Archard proposed that K represents the probability of forming a wear particle in a given encounter. The constant 1/3 in Eq.(3.15d) now gets included in K. The problem with this concept is that there is no valid reason why only one event in a large number of encounters should be successful in removing a particle. Also the wear coefficients can only be obtained on the basis of experiments and cannot be obtained from any physical model. On the basis of wear coefficients the wear may be considered mild or severe. When wear coefficients are less than 10^{-4} the wear may be called mild while it may be referred to as severe when the coefficient is greater than 10^{-3}.

Real surfaces are covered with oxide films. Adhesive wear in such situations can occur only through oxide films and the actual metal contact will be a fraction α of the real area. It is reasonable to assume the average diameter of the transfer particle at an asperity contact is that of a circular patch of metal contact area. On this basis diameter of the transfer particle is $d\sqrt{\alpha}$. Consider that *all* events result in a transfer particle. As the wear particle is generated per event the sliding distance involved remains d as in Archard's model. It can be shown that the wear rate is

$$\frac{V}{l} = \alpha^{1.5}\left(\frac{W}{H}\right)\left(\frac{1}{3}\right) \tag{3.17}$$

where α is fractional metal contact area within the real area

In other words $\frac{\alpha^{1.5}}{3}$ may be considered the wear coefficient that results from the extent of adhesive contact through oxides assuming *all* metallic contacts lead to a transferred particle. The coefficient now has a physical basis. If for example, α is 0.1, the wear rate in the present model is $1.05 \times 10^{-2} \times A_r$ and so the wear coefficient is 1.05×10^{-2}. Depending on α it can have a similar range to that in Archard's equation but has a different meaning. It may be noted that in this model every metal contact leads to a particle and it is assumed that wear is governed by a transfer mechanism alone. Here again the coefficient can only be determined experimentally. This model is only a plausible suggestion to provide a physical basis to adhesive wear coefficient.

Other models have looked at the wear coefficient in terms of contact mechanics. Greenwood's model discussed in chapter 1 can be used to obtain the extent of plastic and elastic contact based on roughness. If it is assumed that only plastic contacts contribute to adhesive wear, wear coefficients can be estimated [15]. Kato et al [16] also considered that adhesive wear occurs when the deformation is plastic and modelled the transfer process. The elastic contacts are considered to wear by a fatigue wear mechanism. In general terms wear due to asperity interaction may be referred to as sliding wear. Sliding wear can be due to several mechanisms including adhesion. The suggestion of the author is to consider adhesive wear as due to overall fractional adhesive metal contact through oxide films. When asperity deformation is plastic larger extent of metallic contact occurs as opposed to elastic contact. Oxides also can wear by a chemical wear mechanism that will be discussed later. Modelling on the basis of one mechanism or the other amounts to an assumption that the presumed mechanism is rate determining. The problem is difficult to resolve on the basis of available literature. Further complications that can arise in adhesive wear situations include large-scale modifications in near surface structures [17] debris compaction in the contact region [18] and work hardening of the asperities that can lead to abrasive wear [19].

From the discussion in the previous paragraph it is clear that definitive modelling of the wear process is not possible. The model due to Archard has practical utility in terms of experimental observations. Many experiments do show that wear rates are proportional to real area within some limits of operating conditions though the detailed mechanisms involved are complex. It is appropriate to consider Archard's model as an equation that describes sliding wear process in general with an experimentally determined coefficient. The wear coefficient will have a specific value for a given range of operating conditions. In another range of operating conditions a different wear coefficient may characterise the wear behaviour. Such variations arise due to several reasons that include near surface material transformations and oxidation. The temperature rise in contact has an important role to play in these modifications. Transitions of a sudden nature can occur when operating conditions exceed certain limits. A typical example of transitions as observed by Welsh [20] that is often quoted in the literature is shown schematically in Fig. 3.3. This diagram illustrates load dependent transitions and recovery due to changes in surface films and changes in hardness of the steel. Wear coefficients can provide limited predictive capability at least in laboratory machines. Wear coefficients are also useful in comparing wear behaviour of different material combinations for screening purposes.

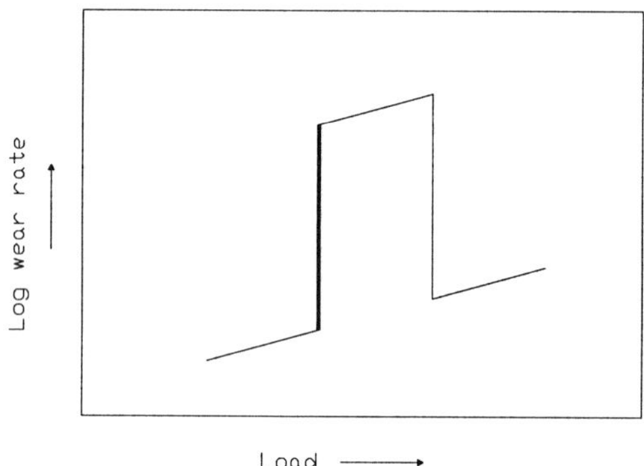

Fig. 3.3 Load wear transitions observed with steel from Ref [20].

As stated earlier sliding wear can have many mechanisms and adhesive wear is one of them. It is possible to study the mechanisms of sliding wear over a wide range of operating conditions and to map the wear behaviour as discussed in the final section.

3.5 Abrasive wear

3.5.1 Phenomena

Abrasive wear is normally associated with material removal when a hard sharp particle slides against a softer material. Grinding of material is a good example of the process in which abrasive grains remove material from the softer surface by a cutting action. In this mode the grain penetrates the surface and the grooved volume is removed during sliding. Abrasive wear may be classified as two-body or three-body. The two-body mode refers to a situation where one of the surfaces is abrasive in relation to the other. Grinding with an abrasive paper is a typical example of this. The three-body mode, as the name suggests, involves abrasion of two surfaces in relative motion due to abrasive third body particles that are drawn between surfaces. Dust particles that get into a lubricating oil circuit can act as third bodies. Wear particles that work harden and circulate between the surfaces can also act as third bodies.

The major portion of research on abrasive wear is with two bodies. This is because the system is better defined in comparison to a three-body system in which the hard particles are ill defined. Early research by Khruschev and Babichev [21] established that for pure metals abrasive wear decreases as hardness increases. Sundararajan [22] found that for pure metals the relative abrasive wear resistance (RAWR) may be expressed as

$\mathrm{RAWR} \propto \mathrm{H}^{0.95}$ where H is the bulk hardness of the metal and RAWR is the relative wear resistance compared to pure iron.

With different steels the change in RAWR with hardness is lower with the exponent varying from 0.2 to 0.55 and is related to the type of steel and its microstructure. This was explained by Sundararajan [23] on the basis of critical strain needed for chip formation.

Cutting action should depend on the relative hardness of the abrasive to that of the metal. Earlier work [3] has shown that abrasive wear becomes negligible when the hardness ratio of the abrasive to the metal is less than the critical range of 1.6 to 1.0.

The above observations refer to situations where cutting wear is predominant. To generalise abrasive action one has to consider particles of varying hardness and shapes moving against a relatively softer surface. If a hard asperity slides over a softer material it can remove material by cutting action or it may only plough a groove without material removal depending on the operating conditions. There can also be wedge formation in the contact region for operating conditions between ploughing and cutting modes. In wedge formation a built up wedge is formed from the softer material. In principle wear mechanisms involved in all these modes may be classified under abrasive wear. The three modes are illustrated in Fig. 3.4 which is based on [24]. It may be noted that plastic flow results in ridges. The ratio β defined as shown in the figure is indicative of the cutting action of the abrasive. In cutting the volume of ridge material is much smaller than the groove volume resulting in $\beta = 1$ for ideal cutting. On the other hand for ideal ploughing $\beta = 0$.

3.5.2 Abrasive wear modelling

The elementary model considers a conical indentor grooving through a softer surface. This is illustrated in Fig. 3.5. A triangular cross section is involved and for ideal cutting the wear volume V will be based on the grooved volume generated over distance l and is given by

$$V = d^2 l \tan\theta \tag{3.18a}$$

where

d		= depth of penetration
l		= distance travelled
θ		= semi-included angle of the cone

Assuming the real area is governed by the hardness and noting that the real area will be half the projected area during motion real contact area is

$$\frac{\pi}{2}(d \tan\theta)^2 = \frac{W}{H} \tag{3.18b}$$

Combining the above two equations (a), and (b)

$$V = \frac{2}{\pi \tan\theta}\left(\frac{Wl}{H}\right) \tag{3.18c}$$

All the grooved material need not be removed as wear volume. This can happen due to several reasons, like repetitive contact in the same groove that changes the contact conditions and the hardness, and abrasive particle size distribution. Ridges as discussed earlier also influence the wear process. Once again there is a need to introduce a wear coefficient K_{ab} as has been done by Archard [25] and the abrasive wear rate may now be expressed as

$$\frac{V}{l} = K_{ab}\frac{W}{H} \tag{3.19}$$

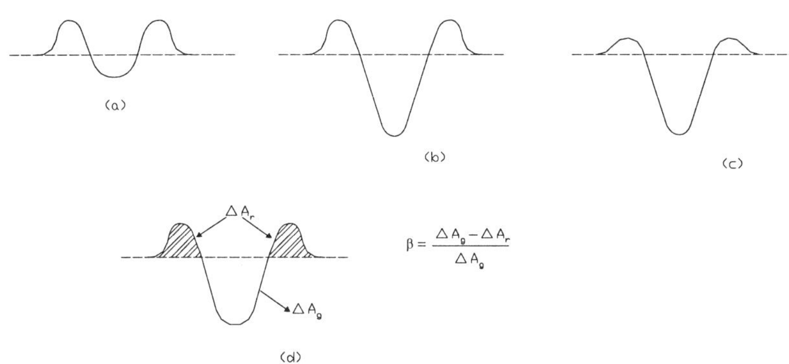

Fig. 3.4. Abrasive wear modes: (a) ploughing (b) wedge action (c) cutting and (d) definition of β based on cross-sectional area of ridge and groove. Based on Ref. [24].

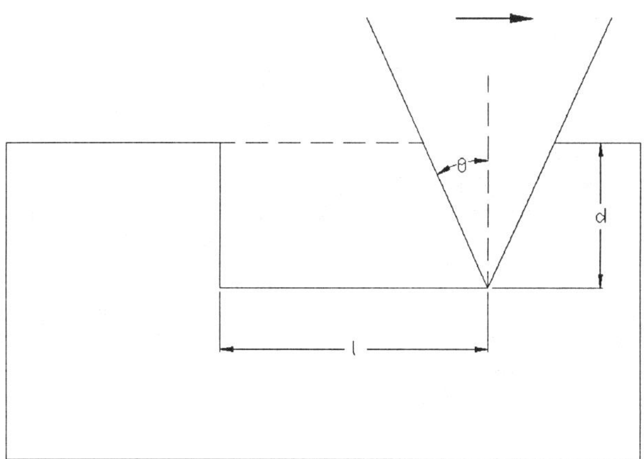

Fig. 3.5. Abrasive cutting by conical indentor.

Abrasive wear coefficients vary from 10^{-1} to 10^{-3} which means the wear rates can be significantly higher than in the case of mild adhesive wear. This is of importance in many lubricated contacts where the deleterious effect of few abrasive particles can outweigh the effective wear protection of lubricants. There has been reasonable success in estimating K_{ab} in a simple scratching mode. Zum Gahr [26] conducted scratching experiments with a diamond pyramid. He estimated β theoretically on the basis of hardness and strain rates and found good correlation between theory and experiment.

The above modelling is valid for ductile materials. Brittle materials in abrasion can undergo lateral cracking due to tensile stresses at the surface leading to brittle fracture. In such cases the problem has to be treated in terms of fracture toughness of the material.

The problem of three-body wear, which is industrially important, has not been modelled adequately. Some qualitative observations can be made on the influence of abrasive particles in a lubricated system. Firstly the particles that induce abrasive wear are those that can enter the contact zone. Thus particles that cause maximum damage are those that are in the same range as the film thickness. In boundary lubricated contacts the influence of particle size distributions and surface hardness on abrasive wear was studied [27]. One interesting observation is the influence of abrasive wear on scuffing. Under some circumstances the large plastic deformation due to abrasive action can induce scuffing at relatively low loads. Many worn surfaces of engineering components show grooving in the sliding direction showing abrasive action. While deeper grooves are due to abrasive action finer grooves may arise from ploughing action. The erosion of materials by impingement of high velocity particles also involves abrasive action.

3.6 Fatigue wear

3.6.1 Phenomena

Fracture and fatigue are well known concepts in material science. It is useful to state these concepts in a simple manner. When a notched bar with a crack length a_c is subjected to tensile stress the stress intensity factor K_I may be expressed as

$K_I = B\sigma(\pi a_c)^{0.5}$ where B is the geometric factor and σ is the applied stress

When a critical stress intensity factor K_{Ic} is reached the material undergoes spontaneous fracture. Thus for a given crack length a stress for fracture is defined. At lower stresses, applied cyclically, the crack grows incrementally finally leading to fracture. This domain is referred to as fatigue. In the case of fatigue the material can undergo large number of stress cycles before failure.

Fatigue wear can occur between surfaces due to cyclic stresses induced at the asperity and sub-surface levels. Such stresses when repeatedly applied can finally result in the detachment of a wear particle that may be termed a micro fracture event. Fatigue wear is more extensively studied in rolling and rolling/sliding contacts under lubricated conditions and is treated in chapter 8. In dry sliding contacts when severe wear is involved transfer wear is likely to dominate. The removal of the transferred material will involve cyclic stressing and fracture but it may be assumed transfer is rate governing. When mild wear is involved the possible oxide removal by fatigue process can compete with the milder transfer process.

3.7 Oxidative wear

3.7.1 Phenomena

Metallic surfaces are covered with oxide films with thickness in the range of 10 to 100 Å. Such films have a protective role in preventing metal transfer as discussed under the section on adhesive wear. The protection from oxide films depends on their adherence to the substrate and their ability to deform without cracking at the asperity stresses involved. Also when locally worn the exposed areas should oxidise and reform the film. This can be considered an ideal situation providing maximum protection against wear. Under severe conditions high surface temperatures lead to further oxidation and thicker films. Such thicker films are expected to wear and reform. This wear mode is referred to as oxidative wear and is confined to the formation and removal of oxide films. This wear mode may also be referred to as film wear as proposed by Archard [25]. In normal air under dry conditions oxygen availability is assured and it is assumed that oxidation is simply governed by the reaction rate. This assumption may be questioned because in the highly stressed asperity contact the available oxygen on each surface can limit the extent of reaction. In the case of lubricated contacts oxygen availability is related to the oxygen solubility in the lubricant and its diffusion in the confined monolayers that exist in the highly stressed asperity contacts. The modelling based on kinetics is of importance not only for oxidative wear but also for chemical wear in which reactions with chemical additives are involved. Oxidative wear can also be treated as chemical wear. In some cases there can be deleterious corrosive wear. One example is the IC engine in which corrosive wear can occur due to acidic components derived from sulphur components in the fuel.

3.7.2 Oxidative wear modelling

First of all the film wear model of Archard [25] is considered. The wear coefficient defined in this equation is then derived on the basis of a kinetic treatment of the oxidation process. The final equation so derived differs somewhat from that of Quinn [28] as explained below.

Consider the earlier wear model of adhesive wear. During asperity encounter let a film of thickness ξ be removed instead of a hemispherical particle. In such a case the wear volume removed at one contact v_d over a sliding distance 'd' will be

$$v_d = \frac{\pi d^2}{4}\xi \tag{3.20a}$$

Following similar argument as per adhesive wear it can be shown

$$\text{Wear rate } V_r = \frac{V}{l} = K_f \frac{\xi}{d}\frac{W}{H} \tag{3.20b}$$

In the above equation K_f is the wear coefficient that is equal to the inverse of the number of encounters needed to form the critical film thickness ξ. It is possible to derive K_f on the basis of chemical kinetics. If t is the total time needed to form a film of critical thickness and t_c is the time during one encounter

$$K_f = \frac{t_c}{t}$$

The time t_c will be d/v where v is the sliding velocity

On the basis of static oxidation tests the oxidation is considered to follow parabolic law. On this basis the increase in film thickness expressed in terms of mass change per unit area Δm, expressed in kg/m^2, will be related to the oxidation time t as

$$\Delta m^2 = k_p t \tag{3.21a}$$

where k_p is the rate constant $(kg^2\ m^{-4}\ s^{-1})$

$$\text{On the basis of kinetics } k_p = A_p \exp\left(\frac{-Q_p}{RT_c}\right) \tag{3.21b}$$

where

A_p = Arrhenius constant, $kg^2\ m^{-4}\ s^{-1}$
Q_p = activation energy, J/mole
T_c = contact temperature in ^0K
R = gas constant, J/mole-^0K

Expressing mass change in terms of density ρ (kg/m^3) of the oxide and taking f as the fraction of the volume that is oxide it can be shown from the above equations

$$K_f = \frac{dA_p \exp\left(\frac{-Q_p}{RT_c}\right)}{\xi^2 \rho^2 f^2 v} \tag{3.21c}$$

Substituting for K_f in Eq. 3.20b

$$\text{Wear rate } \frac{V}{l} = \frac{A_p \exp\left(\frac{-Q_p}{RT_c}\right)}{\xi \rho^2 f^2 v} \frac{W}{H} \tag{3.22}$$

Quinn modelled oxidative wear assuming

Wear rate $V_r = \frac{V}{l} = K\frac{W}{H}$ which is the expression based on adhesive wear in which a hemispherical particle is assumed. His treatment of K in terms of parabolic law is the same as used for K_f here. The present author considers it is more appropriate to model on the basis of film wear as done above. The final expression due to Quinn is obtained by multiplying Eq. (3.20) with d/ξ.

For an oxidation process treating Q_p, A_p, ξ ρ, and H as constants and for simplicity assuming f =1, it can be shown

$$V_r = a_0 \frac{W}{v} \exp\left(\frac{a_2}{T_c}\right) \tag{3.23}$$

where a_0 and a_2 are constants

The importance of this model is to relate K_f to the physical process of oxidation. It should be noted that the oxidation in tribocontacts is different from static oxidation. The rubbing surfaces can have activation energies lower than in static conditions. The assumed parabolic law may or may not be operative. Another issue is the estimation of contact temperatures and the uncertainties involved in such estimations. The significant difficulties involved in reconciling theory with experiment were presented by Quinn [29], who addressed some of the issues raised here. Chemical reactions with antiwear additives can be treated by a similar approach and subject to similar uncertainties.

The oxidative wear model considered above assumes a critical film thickness for oxide removal. This is only an assumption and the exact mechanism of oxide removal need to be understood. The removal mechanism can be fatigue in which case the oxide failure will be related to the oxide structure and composition. These are variables and will not be uniform with changing temperatures. Molgaard [30] has provided a useful discussion of the complexities of oxidative wear. He considered that the mechanical characteristics of the oxide layer are more important than the oxidation itself in the wear process. Also the concept of critical film thickness may only be applicable beyond a particular thickness. Mild wear at low temperatures is governed by tenacious thin films and the present model may not be applicable for such cases.

One possible approach to wear modelling is by designed experiments. A polynomial expression can be developed keeping in view the assumed model. If the model fits then the mechanism may be justified in the range of conditions studied. Further justification is necessary by surface analysis. An example may be cited on the basis of short duration dry wear experiments conducted in a reciprocating tester by Ravi Sankar et al [31]. The objective was to study the influence of load and temperature on the wear rate and whether it involves oxidative wear mechanism. The tests were conducted with cast iron ring piece sliding on a disc cut from a honed liner. Three different loads of 30, 50 and 70 N and three different bulk temperatures of 50, 100 and 150 °C were selected and a 3^2 design adopted. Sliding speed was fixed at 50 Hz with a stroke of 1mm.

If it is assumed that Eq. (3.23) is applicable and v is constant it can be written as

$$\log V_r = a_0 + a_1 \log W + \frac{a_2}{T_c} \tag{3.24}$$

where a_0, a_1 and a_2 are constants. The constant a_1 is included to accommodate possible exponent on W in the empirical relation. The relationships found in this case were

For ring

$$\log V_r = -27.69 + 1.05 \log W - \frac{858.75}{T_c} \tag{3.25a}$$

For disk

$$\log V_r = -30.57 + 1.35 \log W - \frac{750.35}{T_c} \tag{3.25b}$$

The asperity temperature rise was estimated based on section 3.3 and was added to bulk temperature to obtain T_c. In view of the low speeds involved the asperity temperature rise was low and less than 10°C. The role of asperity temperatures can only be established when tested over a wide range. For the test conditions used, the above relationships show that exponential dependence on contact temperature is justified. It is possible that an oxidative wear mechanism is operative here. Flake like removal observed on the liner surface is shown in Fig. 3.6. The surface shown is that observed at a load of 50 N after 25 minutes of running. No analysis of surfaces and wear particles was carried out in this case to prove oxidation is the governing mechanism. The work however shows the importance of approaching wear empirically, and then only to see whether a particular model can explain the data. If the present data cannot be reconciled with the model, other polynomial expressions need to be sought which may or may not fit known models. Wear processes are complex and it is necessary to analyse wear data objectively without a preconceived model. The observed data may partially, and in rare cases fully, justify a known model. These issues will be elaborated upon in later parts of the book.

Fig. 3.6. Worn liner surface showing flake removal.

3.8 Wear map approach

In engineering design the required material properties can be measured by standard tests and directly used. No such methodology exists for wear prediction. Every situation has to be considered case by case and there is a proliferation of test methods. Also the confidence with which predictions can be made is limited. At the fundamental level more sophisticated investigations are revealing information about material transformations, type of oxides, stress distributions and their interactions. Asperity and bulk temperatures play an important role in all these. Thus deeper knowledge is available about specific wear mechanisms. Translating this knowledge into predictive models is the gap in wear research and is very difficult to fill. The unsatisfactory situation regarding wear modelling has been discussed in detail by Ming and Ludema [32]. One possible way is to study over a wide range of operating conditions and map the wear behaviour as a function of operating conditions. Lim et al [33] have studied wear processes with this approach for low carbon steel in a pin-on-disk tester. They divided the sliding wear into zones of ultra mild wear, plasticity dominated wear, mild oxidational wear, severe oxidational wear, melt wear, and seizure. Plasticity dominated wear was modelled considering Archard's model and the model due to Suh and Saka [34]. Model of Suh and Saka is based on the concept of sub-surface plastic deformation leading to fatigue and delamination of wear flakes. This model was not discussed in this chapter. This form of wear typically occurs when there is significant plastic flow at the interface. This wear process also can be reconciled with a wear coefficient as is normally expressed in the Archard's equation. Behaviour in each zone was explained by an available model and approximate equations were provided. The wear map obtained by them is shown in Fig. 3.7 and is a typical example of the approach. Real systems will mainly operate in the mild wear regimes.

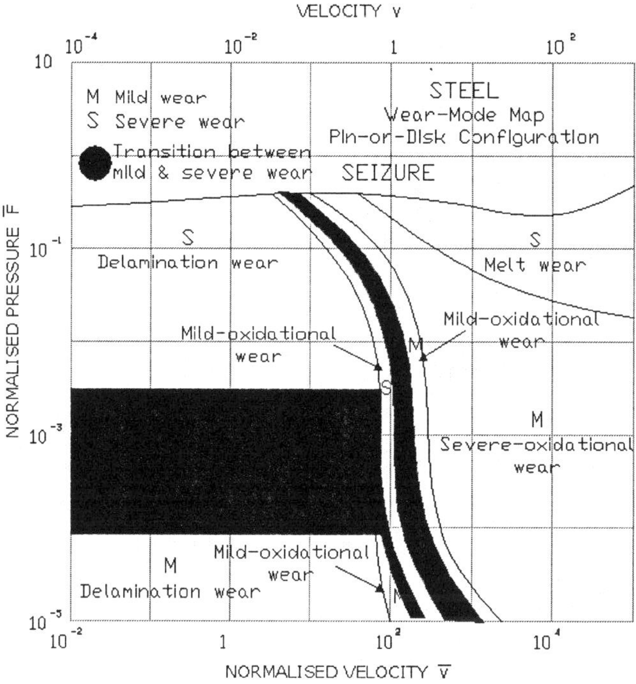

Fig. 3.7. The wear-mode map for unlubricated sliding of mild steel. The various lines separate the regions of dominance. The dark region represents the range in which transition from mild to severe mode occurs. (From Ref. [33] reproduced partially by permission of Elsevier)

In this figure load is in Newtons and sliding velocity is in m/s. The wear rate defined here is the normalised wear rate obtained by dividing the earlier defined wear rate by the geometric area of the pin. Normalised velocity $\overline{v}$ is defined as (velocity x pin radius)/(thermal diffusivity) while normalised pressure $\overline{F}$ is defined as load/(pin area x hardness). Various wear mechanisms observed were due to complex interactions between stress, temperature, and oxidation. The asperity and bulk temperatures were calculated theoretically and had dominating influence on wear mechanisms. These in turn are a function of sliding velocity and normal stress. Such a map can provide the limits of safe operation for a given material from wear point of view. It also provides insight into the wear mechanisms involved. Similar maps are being developed for other materials.

The above approach may be considered semi-empirical and is likely to be developed further in future. Development of such maps is time consuming even with one material pair and one test configuration. The wear maps are expected to change with the test configuration. For example if the tests are conducted in a reciprocating mode or in rolling/sliding contact the results will be different. There can also be variations if the area contact is changed to a concentrated line or point contact. These issues coupled with the need to study a large number of available and potential materials makes the exercise a daunting task. It is also to be noted that when lubricants are used the mechanisms change and different wear maps for such situations are needed. From a practical point of view the range of operating conditions can be narrowed to those expected in service. This will lead to partial wear maps that can be useful. Also in situations of low wear it is necessary to distinguish small variations in wear rates. This is particularly true of lubricated wear. As an example consider the wear coefficient with a given lubricant is 10^{-8}. If this wear rate has to be compared with that of another lubricant, it would be necessary to be able to distinguish variation in wear coefficient of the order of 10^{-9}. This level of distinction is not easy to achieve in experiments. In such cases what is necessary is a detailed mapping in the range of interest with the necessary precision levels. The wear map approach will be further considered in later parts of the book.

References

1. K. H. Zum Gahr, Microstructure and Wear of Materials, Tribology series 10,Elsevier, Amsterdam, 1987.
2. R. G. Bayer, Mechanical Wear Prediction and Prevention, Marcel Dekker, New York, 1994.
3. E. Rabinowicz, Friction and Wear of Materials, John Wiley, New York, 1965.
4. K. L. Johnson, Contact Mechanics, Cambridge University Press, Cambridge, 1985.
5. G. M. L. Gladwell, Contact Problems in the Classical Theory of Elasticity, Sijhoff and Noordhoff, Amsterdam, 1980.

6. J. B. Mann, T. N. Farris and S.Chandrasekar, Effects of friction on contact of transverse ground surfaces, J. Trib.. ASME, 116 (1994) 430.
7. S. C. Lee and N. Ren, The sub-surface stress field created by three-dimensional rough bodies in contact with traction, Trib.Trans., STLE, 37 (1994) 615.
8. X. Tian, F. E. Kennedy, Maximum and average flash temperatures in sliding contacts, J. Trib. ASME Trans., 116 (1994) 167.
9. N. P. Suh, Tribo Physics, Prentice Hall, New Jersey, 1986, 406.
10. O. S. Dinc, C. M. Ettles, S. J. Calabrese and H. A. Scarton, The measurement of surface temperature in dry or lubricated sliding, J. Trib., ASME, 115 (1993) 78.
11. X. Tian and F. E. Kennedy, Contact surface temperature models for finite bodies in dry and lubricated sliding, J. Tribo. ASME Trans., 115 (1993) 411.
12. M. Kerridge, Metal transfer and wear process, Proc. Phys. Soc. 68B (1955) 400.
13. T. Sasada, The role of adhesion in wear of metals, J. Japan Society of Lubrication Engineering, 24 (1979) 700.
14. J. F. Archard, Contact and rubbing of flat surfaces, J. Appl. Phys., 24 (1953) 981.
15. E. P. Finkin, An explanation of the wear of metals, Wear, 35 (1975) 239.
16. T. Kayaba and K. Kato, The adhesive transfer of slip-tongue and wedge, ASLE Trans., 24 (1981) 164.
17. D. A. Rigney, Comments on the sliding wear of metals, Trib. Int., 30 (1997) 377.
18. M. Godet, The third body approach: A mechanical view of wear, Wear, 100 (1984) 437.
19. Y. C. Chiou and K. Kato, Wear mode of micro-cutting in dry sliding friction between steel pairs (Part 1): Effect of attack angle of the specimen,, J. Japan Society of Lubrication Engineering, 9 (1988) 11.
20. N. C. Welsh, Phil. Trans. R. Soc., Ser. A 257, 31 (1965).
21. M. M. Khruschev and M. A. Babichev, Research on Wear of Metals, National Engineering Laboratory, East Kilbridge, chapter 8, 1966, NEL Translation 893.
22. G. Sundararajan, The differential effect of the hardness of metallic materials on their erosion and abrasion resistance, Wear, 162-164 (1993) 773.
23. G. Sundararajan, A new model for two-body abrasive wear based on the localisation of plastic deformation, Wear, 117 (1986) 1.
24. K. Kato, Abrasive wear of materials, Trib. Int., 30 (1997) 333.
25. J. F. Archard, Wear theory and mechanisms, in M. B. Peterson and W. O. Winer (eds.), Wear Control Handbook, ASME, New York, 1980, 35-80.
26. K. H. Zum Gahr, Microstructure and Wear of Materials, Tribology series 10, Elsevier, Amsterdam, 1987, 132-148.
27. S. Odi-Oweri and B. J. Roylance, Lubricated three-body abrasive wear – contaminant condition versus bounding surface material hardness, Trib. Int., 20 (1987) 32.
28. T. F. J. Quinn, The effect of hot-spot temperatures on the unlubricated wear of steel, ASLE Trans., 10 (1967) 158.
29. T. F. J. Quinn, Oxidational wear modelling: I, Wear, 153 (1992) 179.

30. J. Molgaard, A discussion of oxidation, oxide thickness and oxide transfer in wear, Wear, 40 (1976) 277.
31. P. Ravi Sankar, Rajesh Kumar and A. Sethuramiah, Dry wear of grey cast iron, 2000 AIMETA Int. Trib. Conf., Sep 2000, L'Aquila, Italy, 127.
32. H. C. Meng and K. C. Ludema, Wear models and predictive equations: their form and content, Wear, 181-183 (1995) 443.
33. S. C. Lim, M. F. Ashby and J. H. Brunton, Wear-rate transitions and their relationships to wear mechanisms, Acta metall., 35 (1987) 1343.
34. N. P. Suh and N. Saka, The delamination theory of wear, Wear, 44 (1977) 135.

Nomenclature

a	contact radius in point contact
a_a	asperity contact radius
a_c	crack length
a_0, a_1, a_2	constants
A^*	ratio of real area to geometric area of contact
A_g	geometric area of contact
A_p	Arrhenius constant
A_r	real area of contact
b	half contact width in line contact
B	geometric factor
d	average asperity contact diameter
d	depth of penetration of cone
E	effective elastic modulus of the two bodies
E_1, E_2	elastic modulii of the two materials 1 and 2
f	fraction of the volume that is oxide
f	coefficient of friction
$\overline{F}$	normalised load
H	hardness
k'	multiplication factor to obtain maximum shear stress
k_p	rate constant for parabolic growth
$k_{1,2}$	thermal conductivity of material 1 or 2
K	adhesive wear coefficient, non-dimensional
K_{ab}	abrasive wear coefficient, non-dimensional

K_f	film wear coefficient equal to inverse of cycles needed to form critical film thickness
K_I	stress intensity factor
K_{Ic}	critical stress intensity factor
l	sliding distance, m
l^*	ratio of half width of asperity contact to half width of geometric contact
l_a	half width of asperity contact
l_g	half width of geometric contact
p(r)	pressure distribution in circular contact
p(x)	pressure distribution in line contact
p_0	maximum pressure
p_m	mean pressure
P	load
P	load per unit length in line contact
P_e	Peclet number based on geometric contact
P_{e1}, P_{e2}	Peclet number for body 1 or 2 as applicable
q	heat flux
q_a	heat flux on the basis of real area
Q_p	activation energy
R	equivalent radius of two bodies in contact
R	molar gas constant
R_1, R_2	radii of two bodies in contact
t	time needed to form a film of critical thickness
t_c	time during one encounter of asperities
T_c	contact temperature in absolute units
v	sliding velocity
$\bar{v}$	normalised velocity
v_d	wear volume removed at one contact over a sliding distance d
V	wear volume over a sliding distance l
V_d	wear volume generated at all asperity contacts over a sliding distance d
V_r	wear rate, volume/sliding distance
W	load

Greek letters

α	fractional metal contact area within real area
β	$= \frac{\Delta A_g - \Delta A_r}{\Delta A_g}$ as in Fig. 3.4
δ	line contact deformation
Δm	mass change per unit area
$\Delta\theta_a$	asperity temperature rise in contact
$\Delta\theta_{a1}$, $\Delta\theta_{a2}$	asperity temperature rise at the surfaces with all heat flowing to body 1 or 2
$\Delta\theta_{at}$	asperity temperature rise with thermal interaction
$\Delta\theta_{at1}$, $\Delta\theta_{at2}$	asperity temperature rise with thermal interaction with all heat flowing to body 1 or 2
$\Delta\theta_g$	temperature rise in the geometric contact area
ν	Poisson ratio
ν_1, ν_2	Poisson ratios of bodies 1 and 2
θ	semi-included angle of the cone
θ_a	asperity temperature
θ_b	bulk temperature
ρ	average density of oxide
σ	applied stress
τ_{max}	maximum shear stress
ξ	critical film thickness

4. Boundary lubrication mechanisms metallic materials

4.1 Introduction

Lubricated wear occurs at the asperity contacts. In dry contacts the real area is involved in the wear process. In lubricated contacts two different situations arise. In one case the molecular layers adsorbed on the surfaces resist penetration allowing metallic contact only over a small fraction of the real area. This results in substantial reduction in wear. In another situation, which arises under severe operating conditions, adsorbed films cannot survive. In such cases chemical additives are used to generate wear resistant films. This class of additives is referred to as antiwear (AW) additives. Another class of more reactive chemical additives, referred to as extreme pressure (EP) additives, is used to increase the load carrying capacity.

Conventionally boundary lubrication refers to adsorbed films. For the purpose of this chapter the boundary lubrication regime is treated in a broader sense. It is defined as the regime in which molecular layers and/or reacted films provide effective lubrication. Effectiveness refers to reduced wear as well as increased load carrying capacity. The industrial applications of these additives were considered in the second chapter. This chapter is divided into two major sections dealing with adsorbed films and chemical reaction films. Firstly the conceptual physical model for the action of adsorbed films is considered. This is followed by a consolidated treatment of monolayers, and dynamic adsorption followed by a consideration of modern developments. Modern developments include a brief coverage of multilayers and nanotribology. The section on reaction films deals with the main classes of additives and their reactions. This is followed by a consideration of recent developments related to fundamental mechanisms, multi-component additives, and related aspects. Some issues related to the limitations of the available models are raised during the discussion. The final section deals briefly with the commonly used surface analytical techniques.

The aim of this chapter is to provide adequate background for the future chapters dealing with lubricated wear in terms of theory and practice. Some of the recent developments may not be relevant for this purpose but may be of importance in the future. The literature on boundary lubrication is very vast and the coverage is limited to meet the central objective of the book. Detailed consideration of advanced chemistry is not within the scope of this chapter and where necessary only the main reaction paths are mentioned.

4.2 Adsorbed layers

4.2.1 Mechanism of action

It is appropriate to start with the conceptual model of lubrication mechanism due to Bowden and Tabor [1]. In this model the lubricant molecules are considered physically adsorbed on the surfaces. Polar molecules are adsorbed more strongly in comparison to hydrocarbon molecules. The model is usually presented with the polar heads attached to the surface and the hydrocarbon chains projecting vertically upwards. Assume a monolayer of molecules on each surface. When two flat surfaces covered with monolayers are brought into contact under load as shown in Fig. 4.1 the hydrocarbon chains interact preventing direct surface contact. If the surfaces are now slid relative to each other the sliding ideally occurs in the lubricant layers with no metallic contact. In such a situation there will be no wear. Friction also will be low as the shear occurs in the lubricant layers. It may be noted that vertically oriented molecules are 20-30 Å in length and the diagram highly exaggerates the length.

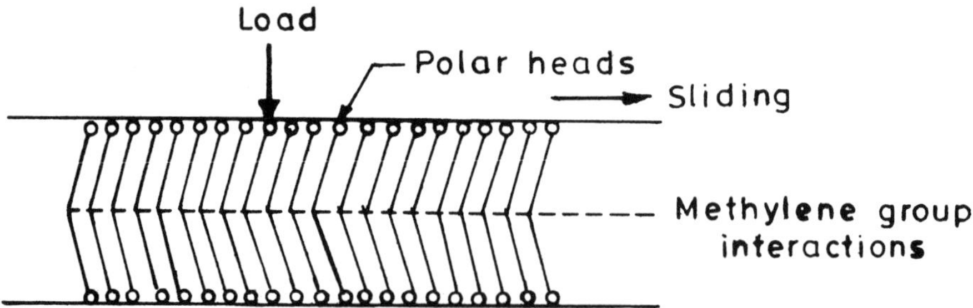

Fig. 4.1. Interacting boundary layers with flat surfaces

The real surfaces are rough and the contact situation for such a case is illustrated in Fig. 4.2a. With reference to the asperity dimensions the molecules will be mere dots. When the surfaces are loaded the load is transmitted to the asperities via the lubricant films. For illustration the boundary layer is shown as a thin continuous film. The stress is assumed to be the same as in the case of dry contacts. Thus the films are subject to high pressures of the order of the metal hardness and the real area A_r is the same as in the case of dry contact. During sliding the shearing occurs in the lubricant layers and ideally no wear should occur. In reality the films are highly stressed and develop defects where metal contact occurs leading to wear. The metal contact is characterised by fractional film defect, α which is defined as the fraction of real area that is metallic. The word 'metallic' needs to be qualified. In real situation the contacts involve oxide as well as metal. In this simple model the contacts are assumed to be metallic. The contact situation with defects is illustrated in Fig. 4.2b that shows a magnified view of contact at the asperity. The flattened asperity is shown having a molecular level roughness and the metal contact occurs at some of the peaks. The overall friction now arises from metallic contact as well as the shear in the film. If shear strength of the film is s_l and the shear strength of the metallic junctions is s_m it can be shown that friction force F is

$$F = A_r[\alpha\, s_m + (1-\alpha)s_l] \tag{4.1a}$$

where A_r is the real area of contact. If now overall friction coefficient is defined by f

$$f = \frac{F}{W} = \alpha\, f_m + (1-\alpha)f_l \tag{4.1b}$$

where

W	= load
f_m	= friction coefficient for metallic junctions
f_l	= friction coefficient for boundary film

In principle for a measured value of f, the fractional film defect can be calculated if f_m and f_l are known. For low value of α, which is the normal zone of interest, variations in f will be too small to be measured. Also f_l and f_m can have variations depending on the operating conditions. Hence the above equation is inadequate to determine α quantitatively.

The above model shows that the efficacy of a boundary layer is governed by the film defect α. The metallic contact is shown to occur at some fine protrusions for illustration. The exact spots where metal contacts occur are not clearly defined in the literature. The fractional metal contact area is normally taken as the fractional area of defects in the adsorbed film. The penetration through the film is related to the tenacity with which the molecules are attached to the surface. This tenacity, or the strength of adsorption, is governed by the nature of the lubricant as well as the material. The actual orientation of the molecules on the surfaces during rubbing process can be very different from the vertical orientation observed in preformed monolayers. In many cases the molecules may lie flat on the surface and the interaction can be over the whole molecular chain during sliding. Another interesting factor is that the friction coefficient of the adsorbed molecules is nearly constant over a wide range of asperity stresses. Such varying stresses can be obtained at the asperities by changing from soft to hard metals. This can be explained if it is assumed that the shear stress increases linearly with pressure as originally suggested by Bridgman [2].

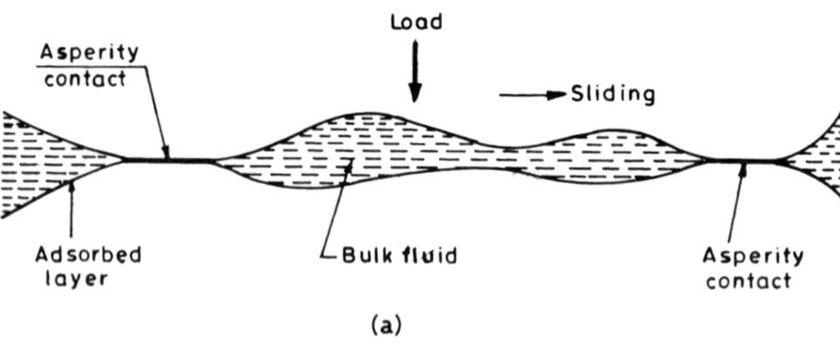

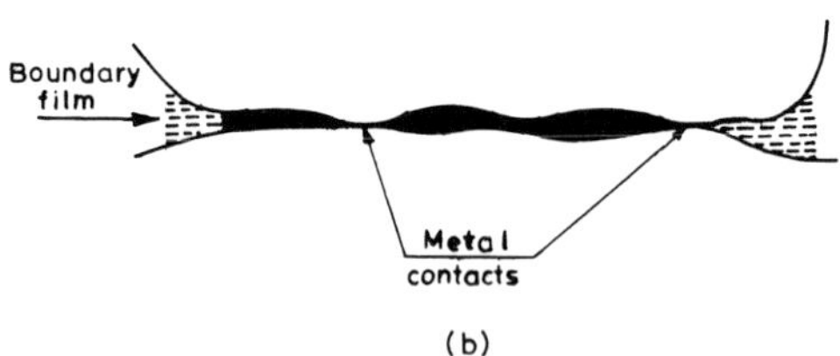

Fig. 4.2. (a) Interacting boundary layers with real surfaces and (b) Detail of metal contact at asperity due to film defect.

The adsorption of the molecules may be physical or chemical. Adsorption involving chemical interaction is referred to as chemisorption and involves chemical bonding with the surface. Several studies were conducted with monolayers to elucidate the detailed mechanism of boundary lubrication. Most of these studies were conducted with preformed monolayers under low speed and low load conditions. The sliding speeds involved were usually less than 10 mm/sec while the loads involved were less than 100g. In these studies the actual orientation of the layers on the surface during rubbing was not known. Generally accepted ideas related to boundary lubrication, and available experimental evidence is presented in the following sub-sections. The presentation aims to provide a consolidated picture of boundary lubrication. Some issues that are controversial in nature are also raised during discussion.

4.2.2 Monolayers as boundary lubricants

Monomolecular films can be deposited by known techniques like transfer of Langmuir-Blodgett films, retraction from melts, and deposition from solutions. Initial pioneering studies were due to Hardy. Later investigations of significance were conducted by Frewing, Bowden and Tabor, Greenhill, Cameron, Zisman, and others. A detailed coverage of the work by all these authors is available in a book [3] published in 1969 that appraised the world literature on boundary lubrication.

The detailed structure of a monolayer of stearic acid from [4] is shown in Fig. 4.3. The stearic acid molecules are vertically aligned with their polar heads anchored to the surface. The chain has a length of 24.2 Å and a cross sectional area of 18.4 Å. This ideal structure may be somewhat modified depending on the nature of surfaces and the techniques used to deposit the monolayer. Such a film can be regarded as solid like. These structures occur with many polar compounds that include alcohols, amines, and fatty acids. Polar molecules are those molecules in which there is a charge separation. This leads to stronger adsorption via the positive polar head of the molecule. Thus the (OH), (NH_2), and (COOH) groups in the above molecules are responsible for the stronger adsorption. The adsorption strength depends not only on the polar group but also on the cohesion between molecules via methylene group interactions.

4.2.2.1 Chain length effects

An effective condensed film can be formed only when chain length is adequate. This has been put in evidence by two different studies. In one study Levine and Zisman [5] deposited monolayers of fatty acids, amines, and alcohols with different chain lengths on stainless steel and glass surfaces. They then measured the contact angle of methylene iodide on the monolayer. Contact angle is a measure of the surface energy. When a surface forms a good condensed film the surface energy is low and the contact angle is high. These measurements showed that the contact angle increased with the chain length and reached a value of ≅ 69° for a carbon chain length of N = 15. The contact angle remained nearly constant for N values in the range of 15 to 26. Below N = 15 there were minor differences between the different compounds with regard to coverage.

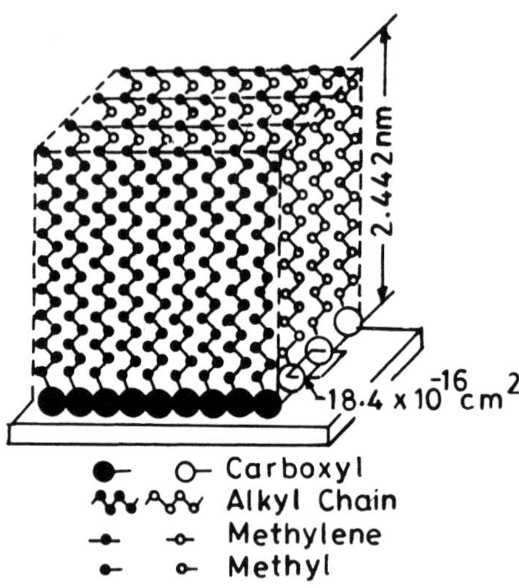

Fig. 4.3. Schematic representation of stearic acid monolayer. (Reproduced from Ref. [4] p204 with permission of Dr. K. C. Ludema.)

Another approach to investigation of the influence of chain length is to study the friction and durability of monolayers. Levine and Zisman [6] have studied the durability of fatty acid monolayers in slow speed sliding. The number of passes needed to obtain sudden transition in friction was taken as the criterion for durability. These studies showed that the long chain fatty acids are more effective and support the findings in the previous paragraph.

Monolayers with branched chains have been found to be less effective. This is attributed to the side chains that decrease intermolecular cohesion. One example is the study conducted with stearic and isostearic acid dissolved in liquid paraffin [7]. Stearic acid showed lower friction in comparison to the isomer when studied as a function of acid concentration in a ball-on-flat steel contact. Though monolayers were not studied it is reasonable to attribute the variation to the nature of the adsorbed acid. Studies by several other authors also support the above findings.

4.2.2.2 Adsorption Vs chemisorption

Physical adsorption is reversible and the molecules are expected to desorb at the melting temperature of the polar compound. Bowden et al [8] studied the influence of temperature on lubrication failure. In these tests 1% lauric acid solution in liquid paraffin was used as the lubricant. In the slow speed test machine the temperature was progressively increased till failure. The lubrication was assumed to be governed by the lauric acid, which adsorbs preferentially on the surface. Several metals were studied in these experiments. The interesting observation was that with unreactive metals the failure temperature ranged from 20°C to 50°C while for the reactive metals it ranged from 90°C to 100°C. The melting point of lauric acid is 49°C and in the case of unreactive metals the failure is governed by the melting point of lauric acid. In the case of reactive metals the failure is attributable to the melting of the chemisorbed soap. The metals that reacted were zinc, cadmium, and copper while the unreactive metals were aluminium, chromium, iron, and silver. Further research in the fifties indicated that even with reactive metals the presence of oxides and humidity are necessary for soap formation [9,10]. Another related aspect is the role of fresh metal surfaces in chemisorption. Earlier work due to Tingle [11] suggested that freshly cut metal surfaces do not promote chemisorption of fatty acids. On the other hand the work of Mori et al [12] showed that organic acids adsorbed strongly on freshly cut aluminium surface under high vacuum conditions. At this stage the contradictions are only pointed out and further consideration will be given to these issues later in the chapter.

4.2.3 Dynamic adsorption and real systems

Studies on preformed monolayers clearly demonstrated their effectiveness in boundary lubrication. In real systems polar compounds are normally used as additives in base fluids. The polar molecules compete with the carrier fluid for adsorption on the surface. A simplified approach to the problem is to consider that the polar molecules dominate adsorption. Assuming Langmuir adsorption it can be shown that the extent of surface coverage is governed by the following equation

$$\frac{\varphi}{c(1-\varphi)} = K \qquad (4.2)$$

where

K = equilibrium constant
φ = fractional coverage
c = concentration of the polar substance in the base fluid

The equilibrium constant is based on the rate constants for adsorption and desorption at the surface. The relationship shows the dynamic nature of adsorption at the surface.

The fundamental thermodynamic equation that governs adsorption is

$$\Delta G^{\circ} = \Delta H^{\circ} - T\,\Delta S^{\circ} \tag{4.3}$$

where

ΔG° = free energy change, J/mol
ΔH° = heat of adsorption, J/mol
ΔS° = entropy change, J/mol
T = absolute temperature, °K

Also

$$\Delta G^{\circ} = -RT(\ln K) \tag{4.4}$$

Combining Eqns. 4.2 to 4.4 it can be shown

$$\ln c = \left[\frac{\Delta H^{\circ}}{RT} - \frac{\Delta S^{\circ}}{R}\right] + \ln\left[\frac{\varphi}{1-\varphi}\right] \tag{4.5}$$

The above has been verified indirectly. The experimental approach involved measurement of failure temperature as a function of additive concentration in the carrier fluid. It was assumed that ΔH° and ΔS° were independent of temperature. Also failure was considered to occur at a particular level of surface coverage φ_c for all additives. Defining T_c as the failure temperature Eq. (4.5) can be recast as

$$\ln c = \frac{\Delta H^{\circ}}{RT_c} + \text{constant} \tag{4.6}$$

Thus $\ln c$ and $1/T_c$ are linearly related at failure. Reasonable linear relationships were found for different fatty acids and soaps by Frewing [13] using a low speed reciprocating tester. Crew and Cameron [14] also found reasonable linear relationships for lauric acid dissolved in different normal alkanes. The implicit assumption here is that the fractional metal contact area is equivalent to (1- ϕ) based on dynamic adsorption. The above equations are as derived in Ref. [13].

In reality the adsorption occurs for both the carrier fluid and the additive. Dorinson and Ludema [15] developed a more complex relationship considering co-adsorption and free energy changes for both solvent and solute. There is also some evidence that improved boundary lubrication occurs when the chain length of the additive and the solvent are matched [16].

Another approach is to measure directly the heat of adsorption and relate the measured value to tribological performance. One important contribution in this area is due to Groszek [17]. He measured the small heat effects due to adsorption by the flow microcalorimeter technique. In this method additive was injected into the flowing solvent and its adsorption on porous adsorbent was measured by the increase in temperature. The heat of adsorption was with reference to the selected flow conditions and the amount of additive injected. The surface coverage estimated in these experiments was less than a monolayer. The important aspect of these studies is the influence of adsorbent besides the known influence of the additive. For example heat of adsorption of stearic acid was found to be 50.8, 13.0, 13.4, and 45.7 cal/g with alumina, silica, carbon, and iron powders respectively. These results were obtained with n-heptane as the solvent. Wear studies with a 4-Ball machine with mineral oil as the carrier fluid showed that wear was reasonably correlated to the heat of adsorption of the additives from benzene on iron powder. Another study in this direction by Hironaka et al [18] showed that the heat of adsorption of stearic acid was much higher on FeS and Fe_3O_4 as compared to Fe_2O_3. High heats of adsorption were very likely due to chemisorption. From a fundamental point of view it is obvious that in real situations where surface oxides with different compositions and morphology exist heats of adsorption cannot be assigned any specific value. Also when chemisorption is involved the desorption will be less important and the mechanism may be simply governed by the soap or other chemisorbed films. Hence the author considers that development of quantitative relationships between wear and heat of adsorption is difficult. However it may be accepted that directionally higher heat of adsorption is beneficial to boundary lubrication.

The models considered above rely on the overall adsorption/desorption on the surface. The contact conditions at the asperities are not invoked. To effectively model the behaviour, asperity temperature and the contact time have to be taken into account. Rowe [19] has proposed a model for wear based on fractional film defect. Firstly the mean time of stay of a molecule at a surface site, t_r can be expressed as

$$t_r = t_0 e^{E/RT_s} \tag{4.7}$$

where

E	= heat of adsorption, J/mol
T_s	= temperature of the surface, °K
t_0	= fundamental time of vibration of the molecule, s
R	= molar gas constant

The probability of desorption of a molecule depends on the time available for it during asperity contact. This time is related to the sliding velocity in the contact. With these considerations Kingsbury [20] proposed the following expression for α the fractional defect

$$\alpha = 1 - \exp\left(-\frac{t_x}{t_r}\right) \tag{4.8a}$$

The time t_x is the time needed to move one molecular diameter and is expressed as

$$t_x = \frac{X}{u} \tag{4.8b}$$

where

X	= diameter associated with an adsorbed molecule, m
u	= sliding velocity, m/s

Combining Eqs. 4.7, 4.8a, and 4.8b the fractional film defect can be expressed as

$$\alpha = 1 - \exp\left[-\frac{X}{ut_0} e^{-E/RT_s}\right] \tag{4.9}$$

For small values of α, less than 0.01, the equation may be simplified to

$$\alpha = \frac{x}{ut_0} e^{-E/RT_s} \tag{4.10}$$

For a given molecular species x and t_0 may be taken as fixed. From the boundary mechanism point of view the calculation of fractional film defect is based on dynamic adsorption coupled with the time needed to move a distance of a molecular diameter. The model thus approaches the real situation and is a step forward in modelling. The wear model based on α shall be considered in chapter 5. The model considered is for a single molecular species. In principle the idea can be extended to multiple additive system by modelling the more complex co-adsorption of the molecules on the surface.

Another idea indirectly related to adsorption is the concept of critical failure temperature. Blok [21] proposed that mineral oils fail at a critical temperature of around 150°C leading to scuffing. Applied scientists and engineers have intensively investigated the validity of this concept. The aspect of lubricant failure will be considered in later chapters.

There is a problem in reconciling the work with monolayers, dynamic adsorption studies, and the concept of critical temperature. For example, long chain paraffin with a melting point of say 40°C should fail to lubricate at this temperature when applied as a monolayer. However such long chain molecules can lubricate effectively to at least 100°C in practical situations. This may be because in normal situations the lubrication is governed by adsorption/desorption while in the case of a monolayer readsorption will be restricted. Polar impurities can also affect the boundary lubrication of hydrocarbons.

4.2.4 Recent investigations of adsorbed films

Recent developments can be divided into two categories. The first category of studies involves better understanding of mechanisms with more sophisticated analytical tools. The studies under ultra high vacuum to elucidate mechanisms form a part of this category. In these studies the tribological contacts involved are described as 'macro' or 'micro'. The distinction between micro and macro contacts is not clearly defined. Even when the contact is macro the asperity contacts are micro in nature and boundary lubrication is governed by the micro contacts. Hence this category is treated as macro/micro contacts. The second category of studies involves elucidation of mechanisms at atomic level. Major investigations are being carried out in this area that is now known as nanotribolgy. The purpose of this section is to highlight the major findings in these categories.

4.2.4.1 Macro/micro level investigations

4.2.4.1.1 The role of oxides

Hu et al [22] have studied the tribochemical reaction of stearic acid on copper surface. Surface enhanced Raman spectroscopy was used to identify the reaction products. Chemisorption was observed when stearic acid was adsorbed in air atmosphere and not in argon atmosphere. This was attributed to the presence of cuprous oxide in air. Cuprous oxide was also detected by Raman spectroscopy. Tribological studies were conducted in a pin-on-disk machine. The layers formed were analysed and copper stearate was detected after running for 400 meters at a load of 15N. The complex reaction route involves initial chemisorption forming a unidentate that eventually gets converted to cupric stearate. These detailed studies were made possible by Raman spectroscopy that can detect low frequency vibrations effectively.

Another interesting study in this direction is an electrochemical study of the interaction between fatty acid and copper oxide by Su [23]. Commercial wire drawing emulsion in water was used for the studies. The variations in current density in an electrochemical cell formed the basis of investigation. Current density increases when oxidation occurs on the scratched copper surface. When stearic acid was injected over copper oxide current density increased only when there is chemisorption. With this technique the author successfully demonstrated that chemisorption occurs with cuprous oxide and not with cupric oxide. It was also observed that a minimum oxide thickness of 60 Å is necessary for effective chemisorption. The author also observed that the fatty acid might be ineffective in wire drawing as formation of the required thickness of cuprous oxide is unlikely. The oxide thickness and its nature were found by electron spectroscopy for chemical analysis (ESCA) and Auger electron spectroscopy (AES). These techniques are considered at the end of the chapter.

The above studies and the earlier studies referred to in the sub-section 4.2.2.2 establish the importance of oxides in chemisorption. Nature of oxide and its thickness are also important. Detailed studies of a similar nature with different metallic oxides can be valuable in selecting polar additives. Such studies have to be coupled with an understanding of the oxides formed in real situations.

4.2.4.1.2 Triboemission

Triboemission consists of exoelectrons, positive ions, and photons. Major interest is in exoelectron emission that occurs during rubbing of surfaces. Such emission can activate the surface and can lead to chemisorption at the surfaces. Significant amount of investigation is being done in this area. The most recent work with diamond pin rubbing against alumina, sapphire, and aluminium [24] showed that major emissions occur with alumina and sapphire. The tests were conducted under a vacuum of 10^{-8} torr. Aluminium surface was less active. Several other research papers can be cited in this area [25,26,27]. Nakayama et al [26] observed that under oxygen pressure of 3×10^{-2} Pa significant electron emission could occur with aluminum surface. This is attributed to the chemoemission that is a consequence of surface oxidation during cutting experiment. With a series of metals they found the chemoemission varied with the ease of formation of the oxide characterised by the free energy of formation. In general the electron emission is favoured by oxidised surfaces. With fresh metal surfaces the emission seems to depend on the extent of plastic straining and fracture. However electron emission can also be a consequence of surface interaction with surrounding species as observed above. It thus appears chemisorption under cutting conditions may or may not occur depending on the operating conditions and environment. This may be the reason for different observations regarding chemisorption of polar molecules in the earlier literature under cutting conditions [11,12]. Exoemission is also of importance in chemical reaction film formation and shall be considered later.

4.2.4.1.3 Ultrahigh vacuum studies

Ultrahigh vacuum can provide an environment where clean surfaces without contaminants can be studied. Friction studies can be conducted in such environment. Lubricant molecules can be introduced into such system and their influence on friction studied as a function of coverage. This leads to precise understanding of the lubrication mechanisms involved. Such studies are being now conducted at an advanced level. Several papers have been published by McFadden and Gellman and [28] is a typical example of this work. While ultrahigh vacuum studies were done in the past the present studies were conducted with instrumentation that can measure adsorption coverage precisely and also ensure initial uncontaminated surfaces. The important observation with pure copper surfaces is that lubrication is ineffective when surface coverage is less than one monolayer. A schematic diagram showing the influence of surface coverage on friction from [28] is given in Fig. 4.4. The studies have so far been mainly conducted with lower molecular weight alcohols. The static friction approaches limiting values at a coverage of 8-10 monolayers. Further studies are likely to clarify fundamental mechanisms involved in boundary lubrication. It may be noted that the tests were conducted at a low temperature of 100°K while the normal force applied ranged from 10-80 mN. The sliding speeds ranged from 0.5-200 μm/s. The static friction coefficients observed even with several monolayers is higher than 0.25 for the alcohols studied. The clean copper surfaces showed friction coefficient in the range of 6.0-7.0. Such high values are expected with very clean surfaces due to junction growth as discussed in chapter 1.

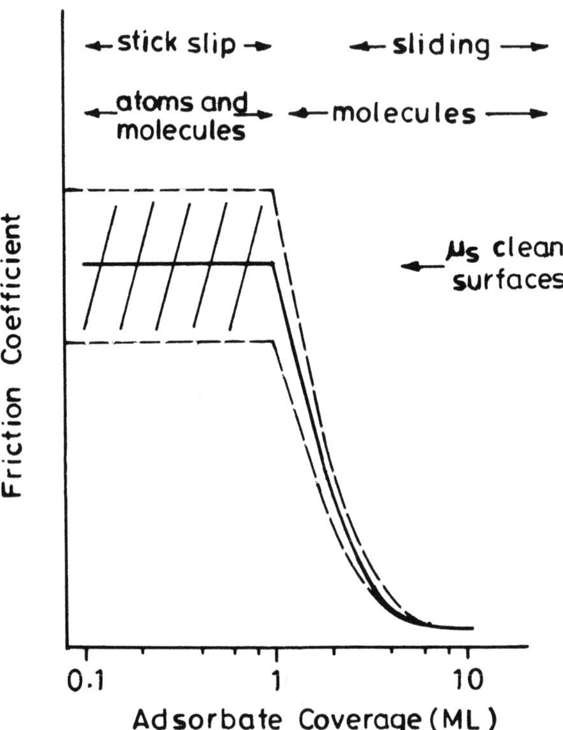

Fig. 4.4. Schematic diagram showing variation of friction coefficient between Cu (III) surfaces with surface coverage. Dashed lines show standard deviation.
(Reproduced from Ref. [28].)

4.2.4.1.4 Multilayers in boundary lubrication

In a sliding or rolling/sliding contact lubricant molecules enter the asperity contact. Several layers of molecules can enter the contact zone. These multlayers can get squeezed out leaving only the tenacious monolayer effective in boundary lubrication. It is also possible that multilayers survive leading to a different form of boundary lubrication. To study such phenomena it is necessary to measure the film thickness as well as the tribological response of the layers. Multilayers involved will be of the order of 100-200 Å. With normal smooth surfaces the asperity dimensions are typically 0.1 μm (1000 Å) and are not suitable to study at the required nanolevel separations. One major approach to such studies has been the surface force apparatus (SFA). In this apparatus molecularly smooth cleaved mica surfaces are used for study. The molecular separations can be measured to an accuracy of 0.1 nm while the force can be measured to a sensitivity of 10^{-8} N. The mica surfaces are curved and the crossed cylinder contact leads to a circular Hertzian contact whose dimensions depend on the normal load. Friction force in sliding can also be measured. The detailed description of the apparatus is available in the book by Israelachvili [29]. Israelachvili and others have conducted several studies with this apparatus and two representative papers in this area may be cited [30,31]. These studies showed that at separations approximately less than ten molecular layers the fluid behaviour could not be handled by the usual continuum theories. Such layers can have viscosity much higher than the bulk fluid viscosity and during sliding can exhibit solid or liquid like behaviour. With shearing the molecular structures can undergo transitions between solid and liquid states leading in some cases to stick-slip phenomena. It may now be argued that boundary layer cannot be treated as a monolayer. However the knowledge base is built around very smooth surfaces with low loads and sliding speeds in the μm/s range. Hence at this stage it is difficult to make any definitive statement with regard to the behaviour of boundary films in normal engineering situations.

Another type of study involves the measurement of film thickness to nanometer level in rolling contact [32]. These studies showed that hexadecane film thickness could be reconciled with the existing theory for EHD. When fatty acids at a small concentration of 0.1% by weight were introduced as additives the film thickness increased by 2-3 times at the lower speeds. At higher speeds exceeding 0.1 m/s the film thickness corresponds to that of hexadecane alone. Another factor brought into focus was the influence of water. When water was present the film thickness at lower speeds increased by 7-8 times and was attributed to the formation of metal carboxylate in the presence of water. These studies also demonstrate that near surface effects leading to viscosity changes can occur with adsorbed species. These studies are yet to be coupled with the studies under sliding conditions in the previous paragraph.

4.2.4.2 Nano level investigations

In SFA a macro contact area of 10-30 μm^2 is involved and the interaction forces are measurable with the apparatus. On the other hand the forces involved in atomic interactions are of the order of 10^{-9} to 10^{-11} N. A far more sensitive apparatus, the atomic force microscope (AFM), has been developed for these studies and is described in detail in the literature [33]. In this apparatus a fine tip approaching the surface can study atomic interactions. The apparatus can also be adapted to measure friction force as well as surface topography. This is achieved by the scanning mode in which the lower surface moves relative to the tip. The control of displacement to within 0.1 Å and the ability to measure very small forces with fine cantilevers are the major features of this apparatus. Development of fine tips with nano scale radii is another important requirement for these studies.

AFM has opened up major investigations in nanotribology. The developments involved are reviewed in some of the recent papers [34,35]. Detailed consideration of this area is outside the scope of the present book. Only major findings are briefly summarised below

1. The adhesive forces involved during approach and retraction normally show hysterisis for solid surfaces. This means that adhesion phenomena are normally irreversible. When lubricant molecules are applied to the surfaces the adhesive forces reduce significantly [36]. In this work it was shown that adhesive forces decreased substantially when stearic acid was applied to an aluminium surface covered with natural oxide.
2. Friction at the atomic scale can be several times lower than for macro contacts [37]. The possible reasons cited include ploughing and loose debris effects in macrocontacts. No explanation seems to be available why static friction itself is high for macrocontacts as compared to nanocontacts.
3. Friction cannot be explained quantitatively and several approaches are being considered. With specific reference to boundary films Israelichvili [38] observed that hysterisis observed in normal approach and retraction mode may be correlated with friction observed in the contact. This reference provides state of the art coverage of molecular films in tribology.
4. Capillary condensation of water between the tip and sample can lead to strong adhesive force due to meniscus formation. This can cause high friction at the start of sliding and is referred to as stiction. Stiction is of

relevance at the head-disk interface in computers. This area has been well investigated and led to the development of perfluoro polyether lubricant coatings grafted to the disk surface. Nanotribological studies have been useful in this development [39,40].

Other investigations being conducted in this area include different coatings, and characterisation of wear and roughness at the nano level. This research is finding application in the computer industry and in micro elecro mechanical systems (MEMS). The applicability of nano level studies to macro contacts is not clear. One approach is to conduct nano level studies on real surfaces and see their applicability to practice. A recent example is the surface obtained in the macroscopic reciprocating tester using molybdenum based additive. This surface was then studied in a force microscope observing both topographical changes and friction [41] along the surface. This study showed the detailed local variations on the surface. Such studies can be an additional tool for understanding the mechanisms involved. In yet another study a more rugged version of force microscope was used to characterise the influence of friction modifiers using actual wet plate clutch materials [42]. The authors found that the traction behaviour observed can be correlated with practice despite large difference in the range of sliding velocities used in the two cases.

4.3 Boundary lubrication with reaction films

4.3.1 Role of chemical additives

Polar molecules considered in the previous sections have temperature limitation of about 150°C beyond which they are ineffective. Many real systems exceed these temperatures and have to be protected by other means. Chemical additives that react with the surfaces and form protective films have been found to be effective for this purpose. Such additives are designated as extreme pressure (EP) additives. A tribological system may operate under varying conditions and the purpose of the EP additive is to act only when needed. Higher temperatures leading to lubricant failure and seizure can be a consequence of high loads, high speeds, or a combination of both. In practical terms the role of additive amounts to increasing the operating severity of the system without failure. Any excessive reaction of the additive will lead to corrosive wear that is undesirable. Hence it is necessary to tailor the additive to limit its reaction to a desirable level when needed.

Another class of chemical additives reacts with the surfaces forming reaction films under milder operating conditions. These additives are called antiwear (AW) additives. As their name suggests, these additives form films that reduce the wear in comparison to a system without these additives. These additives are usually less active than EP additives. In many AW additives the reactive elements are the same as in EP additives. The structure of these compounds is different from those of the EP additives and this difference leads to lower reactivity. These additives are industrially important and used extensively in engine oils, hydraulic systems, and gear oils.

Many real additive systems can have additives that provide both EP and antiwear function. This may be also achieved through a multifunctional additive that may incorporate active elements, say sulphur and phosphorous, in the same molecule. The distinction between AW and EP additives is not clear-cut when additive combinations are used, as the reaction films formed are far more complex.

The purpose of this part of the chapter is to provide information related to the nature of additives and their mechanism of action. The literature available in this area is voluminous and the presentation will be limited to the generally accepted ideas in a consolidated manner. This will be followed by a consideration of the more recent studies and their implications. Detailed consideration is given to wear mechanisms and modeling in the later chapters.

4.3.2 Nature of additives and their reaction mechanisms

The chemical reaction between surface and additive will first involve chemisorption of the additive on the metal surface. This will be followed by decomposition of the molecule leaving the reactive elements attached to the metal surface. Further reaction leading to formation of a reacted film of a given thickness involves diffusion of the active elements into the surface coupled with migration of metallic ions towards the surface. The ensemble of these reactions can be reasonably studied under static conditions with approaches well known in physical chemistry. The reactivity under static conditions is then related to the tribological performance usually determined with standard test machines. This approach suffers from two deficiencies. The first is the known fact that films formed in dynamic conditions involving wear and reformation will be very different from the static films that involve no rubbing. The second problem is that the performance as judged by a given standard machine cannot be generalised and is at best representative of the performance under the chosen conditions. These aspects were broadly considered in section 2.5.3. These issues need serious consideration and shall be dealt with in the relevant chapters. As detailed consideration of test machines and test procedures shall be taken up later the performance tests used are not treated in detail here. The main approach involved in testing is only mentioned. These issues are raised at the outset so that the reader is aware of the limitations within which EP and AW additive mechanisms are normally studied. The additives are categorised into sulphur, phosphorous, and chlorine compounds so that their mechanisms can be explored. This is followed by a consideration of the mixed systems and multifunctional additives.

4.3.2.1 Sulphur compounds

Large numbers of organic sulphur compounds are used which include elementary sulphur, sulphurised fats, monosulphides, and disulphides. Forbes and co-workers proposed that the reactivity of sulphur compounds is related to the bond energy of the C-S bond [43,44] in the additive molecule. Lower bond energy results in easier reaction with the surface and results in better EP action. Dorinson and Ludema [4] listed the typical sulphur compounds and their structures. This information is reproduced in Table 4.1. Bond dissociation energies for C-S bond are also given in the same table. In the case of elemental sulphur the bond energy refers to the S-S bond. Disulphides in general have lower bond energies than monosulphides and hence are considered to be better EP additives. But these statements have to be qualified in terms of the nature of alkyl and aryl groups involved which have a substantial influence on the bond strength. Thus it may be seen that phenyl methyl sulfide having a bond energy of 335 kJ/mole is much less reactive than benzyl ethyl sulphide with a bond energy of 222 kJ/mole. Similar arguments apply to disulphides and it is known that dibenzyl disulphide reacts more easily than diphenyl disulphide. Dibenzyl disulphide is considered an EP additive while diphenyl disulphide is considered an AW additive. In 4-Ball tests EP properties are characterised by the weld load of balls. Faster reaction results in better protection of the surface and the welding load will be higher. Short duration tests are conducted with a fresh set of balls at each load. Other methods used to characterise EP properties are mean hertz load (MHL), and seizure delay load. Forbes et al [44] reported that bond dissociation energy and EP properties are reasonably correlated based on tests with a series of sulphur compounds.

Sakurai et al [45] characterised the EP properties by a different approach. In this approach firstly chemical reactivity was characterised by static reaction over a hot wire. The film thickness was obtained on the basis of change in electrical resistance of the hot wire. The film grows with time offering a barrier to further reaction. The film thickness growth rate follows a parabolic law for the majority of the chemical additives studied. The relation may be expressed as

$$\Delta r^2 = kt \tag{4.11}$$

where

Δr = film thickness, m
k = rate constant, m^2s^{-1}
t = time, s

Table 4.1. Bond dissociation energies in various organosulphur compounds (Reproduced from Ref. [4] p263 with permission of Dr. K. C. Ludema.)

Bond	Structural type	Dissociation energy, kJ/mole
S_6 ring (S—S)	Elemental sulphur	117-147
RSS—SSR	Dimethyl tetrasulphide	151
RS—SR	Disulphides	280-289
R—SR	Alkyl monosulphides	293-306
iso$-C_3H_7-SCH_3$	sec-Alkyl monosulphides	280
t$-C_4H_9-SCH_3$	t-Alkyl monosulphides	272
$CH_3CH=CH-SCH_3$	Vinyl methyl sulphides	218
$C_6H_5S-CH_3$	Phenyl methyl sulphide	251
$C_6H_5CH_2S-CH_3$	Benzyl methyl sulphide	213
R—SH	Alkyl mercaptans	289-310
iso$-C_3H_7-SH$	sec-Alkyl mercaptans	297
t$-C_4H_9-SH$	t-Alkyl mercaptans	289
$CH_3CH=CH-SH$	Vinyl mercapatans	218
(ring)—SCH_3	Phenyl methyl sulphide	335
$C_6H_5CH_2-SC_2H_5$	Benzyl ethyl sulphide	222

By conducting the studies at different temperatures the rate constant can be obtained at different temperatures. Load carrying capacity was obtained on the basis of mean Hertz load (MHL). Fig. 4.5 from [45] shows reasonable correlation between relative reactivity and load carrying capacity. The relative reactivity is the ratio of k values for additive solution and white oil obtained at 400°C. It may be noted that the reactions were studied with sulphur as well as chlorine compounds. When mixed additives were used the sulphur additive dominated the effect. Some earlier investigations also involved reaction studies with metallic powders and relating the global reaction to the reactivity of the additives. In these studies the nature of decomposition products was also analysed by gas chromatography and mass spectrometry and possible reaction routes postulated.

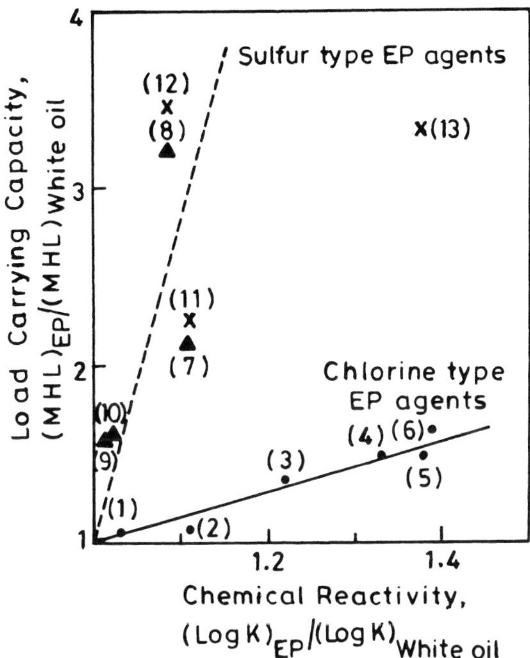

Fig. 4.5. Correlation between chemical reactivity and load–carrying capacity of oils containing EP agents. (▲) sulphur compounds. (•) chlorine compounds. (×) mixed sulphur-chlorine compounds.
(Reproduced from Ref. [45] by permission of STLE)

The reaction films generated on the surfaces have been analysed with increasing levels of sophistication. Early detailed work of Godfrey [46] showed that the films formed with sulphur compounds are composed of both sulphides and oxides. He also showed that preformed sulphide films had lower load carrying capacity in comparison to mixed films formed in air atmosphere. The relative proportion of oxides and sulphides in the film depends on the operating conditions. Also different types of oxides can be involved in the film. Later work by several authors [47,48,49] confirms the presence of oxides in surface films. It is also established that oxides formed increase the load carrying capacity of the sulphur compounds. Thus the reaction films formed are due to simultaneous action of sulphur and oxygen competing for the surface. The hot wire reactions of the previous paragraph should now be qualified. Even for the static reaction the films are ill-defined combinations of sulphides and oxides, though the overall increase in film thickness could be expressed by the typical parabolic law of corrosion reactions.

Additives with lower levels of reactivity are unsuitable as EP additives. Some of them can be used as antiwear additives. The major class of AW additives used are the phosphorous compounds which are considered next.

4.3.2.2 Phosphorous compounds

The phosphorous compounds used are mainly phosphate and phosphite esters. These compounds are manly used as antiwear additives. The antiwear function was evaluated by the scar diameter obtained at 15 kg load in a 4-Ball machine in several earlier papers [50,51]. Based on earlier work there is reasonable evidence that the mechanism of action of these additives involves hydrolysis of the esters followed by reaction with the surfaces. The final wear resistant films consist of phosphate or phosphite films together with oxides. It is also considered that phosphites provide better antiwear protection at higher loads in comparison to the corresponding phosphates. Tricresyl phosphate is the most common phosphorous compound used in this class. In many industrial formulations both sulphur and phosphorous compounds are used together and in such cases it is difficult to define the nature of films formed. In phosphates and phosphites a large number of structures with partial or complete replacement of hydrogen are possible. The alkyl and aryl groups also can be chosen as desired. The typical structures for fully substituted phosphate and phosphite esters are $(RO)_3P=O$ and $(RO)_3P$ respectively. The reactivity in relation to antiwear behaviour is more difficult to define. All that can be said is that the effective additives provide optimum level of reaction to sustain an antiwear film. This is unlike the case with EP additives where higher reactivity gives better EP action.

4.3.2.3 Chlorine compounds

Chlorine compounds can also be used as EP additives. Their use in industry is declining due to environmental considerations and the possibility of rusting of ferrous materials via iron chlorides. Some operations in which they still find application include severe cutting operations using neat cutting fluids, and stainless steel wire drawing. The compounds used are alkyl and aryl chlorides. Correlation of EP activity with C-Cl bond strength was not found to be reliable. On the other hand reactivity on hot wire as shown in Fig. 4.5 correlated well with EP activity. The overall mechanism involves decomposition of the chlorine compound and formation of metallic chloride.

4.3.2.4 Multifunctional additive systems

The individual additives from different classes can be mixed together to obtain desired antiwear and EP performance. Mixing of two boundary lubrication additives can alter the functioning of the individual additives both of which now compete for the surface. This can result in synergism or antagonism of the additives. These issues were briefly considered in chapter 2. Two systems can be envisaged. One is the isolated system containing only the EP and antiwear additives in a base oil. Studies with such systems reported in the earlier literature are difficult to analyse because different criteria were adopted by the various authors for evaluation. Some used the standard tests, while others used modifications of these tests. Some authors found that chlorine and sulphur additives have a synergistic effect with regard to antiwear and EP functions [52,53]. With regard to combinations of sulphur and phosphorous additives it is generally considered that they act synergistically [54].

Additives can be synthesised with two or more active elements in the same molecule. Thus compounds containing chlorine-sulphur, chlorine-phosphorous, sulphur-phosphorous are also available. By tailoring the functional groups carefully required performance equivalent to or better than individual additives can be achieved. In view of the complexity of the functioning of these additives experience in practice has a role to play in selection.

Another important class of additives is the zinc dithiophosphates that are extensively used in engine oils. In view of their importance the mechanism of their action is being studied over a long period and continues to be pursued even now. These additives shall be considered under recent developments.

Many industrial lubricant systems are complex and may contain several other additives like antioxidants, detergents, antirust compounds, and friction modifiers as considered in the second chapter. These additives in turn will influence the EP and AW function of the additives. Further complication arises from the fact that each of these additives can have different structures depending on the requirements. It appears at present the problems are overcome on the basis of empirical information. A rough indication of the influence of various additives may be seen in Table 4.2 reproduced from [55]. A negative sign indicates antagonism while a positive sign indicates synergism. Exclusory effect refers to one additive excluding the effect of the other while complimentary effect refers to the reinforcing action of each other. Graded response refers to the case in which mixtures of additives of a given function are added to provide effective response under different operating conditions. For example a mixture of phosphorous compounds may be used to obtain good AW performance under varied operating conditions. This table is based on the response of the two component systems. With multicomponent systems the situation can be more complex.

Table 4.2. Main additive interactions
(Reproduced from Ref. [55] by permission of Leaf Coppin Publishing Ltd.)

Additive pair	Main types of interaction	
Antioxidant-antioxidant	Complementary response	+
AW/EP-AW/EP	Exclusory effects at surface (EP/EP)	-
	Complementary response (EP/AW, AW/AW)	+
	Graded response (EP/AW, AW/AW)	+
EP-friction modifier	Exclusory effect at surface	-
EP-rust inhibitor	Exclusory effect at surface	-
Dispersant/ZDDP	Complex formation in liquid	-
	Complex formation at surface	+
Detergent/ZDDP	Solubilization of AW film	-
	Direct reaction in liquid phase	-

4.3.3 Recent investigations of reaction films

Recent developments include mechanistic studies at a fundamental level and characterisation of reaction films with sophisticated analytical tools. Such analysis is also being applied to study additive-additive interactions and to understand mechanisms of action with newer classes of additives. Sophisticated chemistry is involved in these studies, which is not discussed here. The coverage only serves to highlight these studies and points to some missing links from the application point of view.

4.3.3.1 Mechanistic studies

Tribochemical reactions are very complex and a specialised book on tribochemistry was published in 1984 by Heincke [56]. The complexity arises from the fact that mechanical activation of the surfaces can lead to a host of processes. These include lattice and grain boundary effects, catalytic activity of metal surfaces, exoelectron emission, and thermal excitation. Several interesting examples of dramatic effects on reaction with mechanical stress have been cited in this book. While directional influences are well documented very little can be said with regard to specific application of these ideas. While exoelectron emission is known [24-27] its role in tribochemistry has been modelled more precisely by Kajdas [57]. His essential argument is that the low energy electrons generated from the surface due to rubbing action interact with the additive molecules generating negative ions. The positive spots on the surface (generated due to the removal of electrons) chemisorb negative ions. The organic part of the chemisorbed molecule can undergo bond scission leaving behind the inorganic layer. The earlier model due to Forbes [43] considered only thermal activation as the source of electron transfer and covalent bonding. Based on [57] the chemisorption mechanism of monosulphide may be explained as below:

Surface rubbing generating low energy electrons

Interaction of the monosulphide with electrons as

$$R - S - R + e \rightarrow RS^- + R^O$$

$\downarrow$

Chemisorption of RS^- on the positively charged metal surface

In the case of disulphide the initial interaction leads to RS^- and RS^O. RS^O in turn can react with a low energy electron to generate another RS^-. Thus increased chemisorption occurs with the disulphide. It is considered that the chemisorbed species decomposes eventually leading to iron sulphide. The fate of organic radicals depends on their nature and environment. The above example is a simple case to illustrate the idea and can be extended to other classes of additives. Reactions can also occur with base fluids and oxygen as discussed in the same reference. Furey et al [59] have utilised the concept to generate friction polymer that can be effective in lubrication. Monomers dissolved in base fluid can polymerise due to reactions initiated by exoelectrons.

One missing link is the mechanism of film growth due to post chemisorption reactions. This involves diffusion of metal ions and active elements. As it is the reaction film that is involved in EP and antiwear action, classification of efficacy only in terms of the initial reactivity is inadequate. The earlier work due to Sakurai [45] where overall reactivity was characterised in terms of film growth is in the right direction. These remarks pertain to static reactions. In dynamic situations the film composition itself is a variable and more difficult to model. The question is what kind of fundamental investigations will help in modelling the real situations. This question will be taken up in later chapters.

4.3.3.2 Characterisation of reaction films

The reaction films formed with chemical additives are complex in nature. These films are being analysed with modern surface analytical tools and there is better understanding of the films now. Only one example of the reactions involved with zinc dithiophosphate (ZDTP) is considered here. The typical structure of ZDTP is

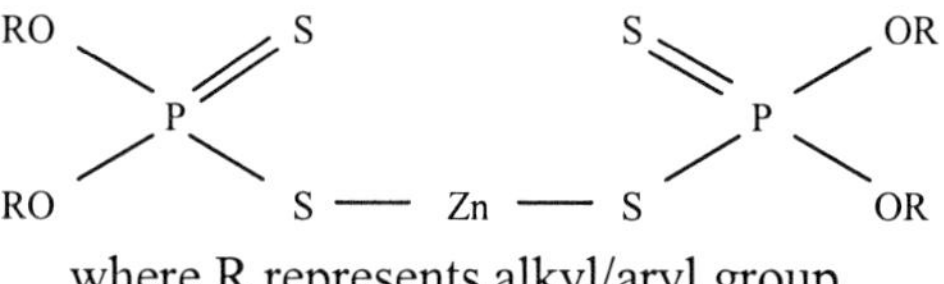

where R represents alkyl/aryl group

Both alkyl and aryl groups can be incorporated in the same molecule providing flexibility with regard to additive response. This class of additives is widely used in engine oils and they are referred to as multifunctional additives. They have antioxidant, antiwear, and a level of EP action. Wilmert et al [59,60] elucidated the nature of films formed in a cam and tappet rig after careful work. The film consists of inorganic amorphous phosphate deposited over the surface. The reaction mechanism postulated involves initial oxidation of the dithiophosphates. The reaction products act as precursors for the formation of the inorganic film. The phosphate chains interact with zinc complexes and the zinc cations get incorporated in the polymer. Detailed analysis of the films is possible through modern tools like secondary ion mass spectroscopy, nuclear magnetic resonance,

Raman spectroscopy, and advanced infrared techniques. To understand reaction mechanisms leading to the final film it is necessary to understand in detail the interactions in the liquid phase also. It is also of interest to note that the authors could also clarify the influence of overbased detergent on the phosphate structure. Such studies are being conducted with other additive systems and in future the detailed film compositions for several combinations may be available [61,62]. The discussion here is only indicative and by no means exhaustive. The references cited are interesting examples of this vast area. Several studies are also being conducted with regard to synergism and antagonism of mixed additive systems where again the sophisticated analytical tools are useful [63,64]. Such interaction studies are of use in formulating oils with the least adverse effects. Yet another interesting approach to investigations of the mechanisms involves tribological studies under ultra high vacuum and analyzing the films in-situ as they form [65].

New classes of additives are being synthesized in large numbers and their performance based on standard tests is being reported. Dithiophosphates where Zn is replaced by other metals, and organic molecules containing N, S, Cl, and P combinations [66,67] are examples of such additives. Yet another class of additives being studied is the boron compounds [68]. The future of these additives in terms of commercial use is yet to be ascertained.

4.4 Surface analysis

Understanding of surface interactions has been made possible by modern analytical tools. These analytical techniques are evolving continuously. This section is limited to a brief consideration of three commonly used techniques in the following paragraphs. Besides surface analysis the modifications that occur in the additives and base fluids are also important. The commonly used techniques include infrared (IR) spectroscopy, mass spectrometry, and nuclear magnetic resonance (NMR) for lubricant analysis. IR techniques are also being applied to study the composition of surface layers by modification of the conventional techniques. Some publications available may be consulted for detailed information [69,70,71].

Electron probe microanalysis (EPMA) is a commonly used technique for elemental analysis. This technique is usually coupled with scanning electron microscopy. In this technique the electrons that impinge on the surface generate the characteristic X-rays of the elements present. The instrument then analyses the energy intensity at the wavelengths that are characteristic to the elements. The technique involves penetration depths of the order of one micron. Semi-quantitative analysis of the concentration of the elements is possible by this technique. The technique is less reliable for elements below carbon and does not provide information on the nature of bonding between the elements.

The X-ray photoelectron spectroscopy (XPS) is another technique being increasingly used in tribology. This technique was earlier known as electron spectroscopy for chemical analysis (ESCA). In this technique the X-rays that impinge on the surface emit electrons at different energy levels characteristic of each element. The instrument provides output giving a plot of binding energy and corresponding intensity. While binding energies are specific to each element, any shift in the binding energy is characterised by the nature of chemical bonding involved. This enables analysis of the nature of compounds formed from the known information of the shifts in reference compounds. Tribological surfaces may have compounds that are not well defined and in many cases clear identification is difficult. The typical depth involved is 100 Å and hence the analysis is representative of the top layers of the surface.

Another technique used for surface analysis of the topmost layers is the Auger electron spectroscopy (AES). The technique is based on the impingement of electrons with specific energy level on the surface. This leads to interaction of electrons in different shells and release of an Auger electron from the M shell. The analysis is based on the energy levels and their intensity. The analysis is limited to few atomic layers and so is representative of the topmost surface layers. The technique is well suited to detect lighter elements. Both XPS and AES are capable of detecting all elements above helium. The AES technique is normally not used for detecting the chemical bonding involved.

In tribological research XPS and AES are also used for depth profiling. In depth profiling the elemental composition is obtained as a function of depth. This is achieved by removing surface layers by ion bombardment at a known rate and analysing the surface at different stages. This information is of importance to understand the variations of composition as a function of depth. The techniques discussed above involve high vacuum and are not suited to analyse organic layers on the surface. For such an analysis modern IR techniques involving multiple reflection are well suited. Certain aspects of surface interaction can be characterised better by Raman spectroscopy which was used in some tribological investigations.

References

1. F.P. Bowden and D. Tabor, The Friction and Lubrication of Solids, Oxford University Press, Part I, 1950.
2. P. W. Bridgman, Rev. Mod. Phys. 18 (1946) 1.
3. F. F. Ling, E. E. Klaus and R. S. Fein (eds.) Boundary Lubrication – An Appraisal of World Literature, ASME, New York, 1969.
4. A. Dorinson and K. C. Ludema, Mechanics and Chemistry in Lubrication, Tribology series 9, Elsevier, (1985).
5. O. Levine and W. A. Zisman, Physical properties of monolayers adsorbed at the solid-air interface-I. Friction and wettability of aliphatic polar compounds and effects of halogenation, J. Phys., Chem. 61 (1957) 1068.
6. O. Levine and W. A. Zisman, Physical properties of monolayers adsorbed at the solid-air interface-II. Mechanical durability of aliphatic polar compounds and effect of halogenation, J. Phys., Chem., 61 (1957) 1188.
7. S. Jahanmir and W. Belter, An adsorption model for friction in boundary lubrication, ASLE Trans., 29 (1986) 423.
8. F. P. Bowden, J. N. Gregory and D. Tabor, Lubrication of metal surfaces with fatty acids, Nature, 156 (1945) 97.
9. H. A. Smith, The adsorption of nanodecanoic acid on metal surfaces, J. Phys. Chem., 58 (1954) 449.
10. W. Hirst and J. Lancaster, Effect of water on the interaction between stearic acid and free powders, Trans. Faraday Soc., 47 (1951) 315.
11. E.D. Tingle, The importance of surface oxide films in the friction and lubrication of metals, Trans. Faraday Soc., 326 (1950) 97.
12. S. Mori, M. Sujinoya and Y. Tamai, Chemisorption of organic compounds on a clean aluminium surface prepared by cutting under high vacuum, ASLE Trans., 25 (1982) 261.
13. I. J. Frewing, The heat of adsorption of long-chain compounds and their effect on boundary lubrication, Proc. Roy. Soc. London, Series A, 182 (1944) 270.
14. W. J. S. Crew and A. Cameron, Thermodynamics of boundary lubrication and scuffing, Proc. Roy. Soc. London, Series A, 327 (1972) 47.
15. A. Dorinson and K. C. Ludema, Mechanics and Chemistry in Lubrication, Tribology series 9, Elsevier, 1985, 229-230.
16. T. C. Askwith, A. Cameron and R. F. Crouch, Chain length of additives in relation to lubricants in thin films and boundary lubrication, Proc. Roy. Soc. London, Series A, 291 (1966) 500.
17. A. J. Groszek, Heat of preferential adsorption of surfactants on porous solids and its relation to wear of sliding steel surfaces, ASLE Trans., 5 (1962) 105.

18. S. Hironaka, Y. Yahagi and T. Sakurai, Heat of adsorption and anti-wear properties of some surface active substances, Bull. Japan Petrol. Inst., 17 (1975) 201.
19. C. N. Rowe, Some aspects of the heat of adsorption in the function of a boundary lubricant, ASLE Trans., 9 (1966) 101.
20. E. P. Kingsbury, Some aspects of thermal desorption of a boundary lubricant, J. Appl. Phys., 29 (1958) 888.
21. H. Blok, The flash temperature concept, Wear, 6 (1963) 483.
22. Z. Hu, S. M. Hsu and P. S. Wang, Tribochemical reaction of stearic acid on copper surface studied by surface enhanced Raman spectroscopy, Trib. Trans., STLE, 35 (1992) 417.
23. Y. Su, Electrochemical study of the interaction between fatty acid and oxidized copper, Trib. Int., 30 (1997) 423.
24. G. J. Molina, M. J. Furey, A. L. Ritter and C. Kajdas, Triboemission from alumina, single crystal sapphire, and aluminium, Wear, 249 (2001) 214.
25. J. Ferrante, Exoelectron emission from a clean annealed magnesium single crystal during oxygen adsorption, ASLE Trans., 20 (1977) 328.
26. K. Nakayama, J. A. Leiva and Y. Enomoto, Chemi-emission of electrons from metal surfaces in the cutting process due to metal/gas interactions, Trib. Int., 28 (1995) 507.
27. K. Nakayama and H. Hashimoto, Triboemission, tribochemical reaction, and friction and wear in ceramics under various n-butane gas pressures, Trib. Int., 29 (1996) 385.
28. C. F. McFadden and A. J. Gellman, Metallic friction: the effect of molecular adsorbates, Surface Science, 409 (1998) 171.
29. J. N. Israelachvili, Intermolecular and Surface Forces, Second edition, Academic Press Limited, London, 1992.
30. V Depalma and N. Tillman, Friction and wear of self-assembled trichlorosilane monolayer films on silicon, Langmuir, 5 (1989) 868.
31. G. Luengo, J. Israelachvili and S. Granick, Generalised effects in confined fluids: new friction map for boundary lubrication, Wear, 200 (1996) 328.
32. M. Ratoi, V. Anghel, C. Bovington and H. A. Spikes, Mechanisms of oiliness additives Trib. Int., 33 (2000) 241.
33. D. A. Bonnell (ed.), Scanning Tunneling Microscopy and Spectroscopy-Theory Techniques and Applications, VCH Publishers, New York, 1991.
34. B. Bhushan, J. N. Israelachvili, U. Landman, Nanotribology: friction, wear and lubrication at the atomic scale, Nature, 374 (1995) 607.
35. B. Bhushan, Nanoscale tribolophysics and tribomechanics, Wear, 225-229 (1999) 465.
36. N. A. Burnham, D. D. Dominguez, R. L. Mowery and R. J. Colton, Probing the surface forces of monolayer films with an atomic-force microscope, Phys. Rev. Lett., 64 (1990) 1931.
37. B. Bhushan and A. V. Kulkarni, Effect of normal load on microscale friction measurements, Thin Solid Films, 278 (1995) 49.

38. A. D. Burman and J. N. Israelachvili, Microtribology and microrheology of molecularly thin liquid films, in B. Bhushan (ed.), The Modern Tribology Handbook, CRC Press, New York, 2000, Section 16.
39. N. Binggeli, R. Christoph, H. E. Hinterman and C. M. Mate, Nanotribology at the solid-liquid interface under controlled conditions, in: B. N. J. Persson and E. Tosati (eds.), Physics of Sliding Friction, Kluwer Publishers, Netherlands, 1995, 415.
40. V. N. Koinkar and B. Bhushan, Micro/nano scale studies of boundary layers of liquid lubricants for magnetic disks, J. Appl. Phys., 79 (1996) 8071.
41. K. T. Miklozic, J. Graham and H. A. Spikes, Characterization of boundary films by atomic force microscopy, 2000 AIMETA Int. Tribology Conf., Sep. 2000, L'Aquila, Italy, 737.
42. C. G. Slough, H. Ohtani, M. P. Everson and D. J. Melotik, The effect of friction modifiers on the low-speed friction characteristics of automatic transmission fluids observed with scanning force microscopy, SAE paper 981099, (1998) 420.
43. E. S. Forbes, The load carrying action of oragano-sulphur compounds – A review, Wear, 15 (1970) 87
44. E. S. Forbes and A. J. D. Reid, Liquid phase adsorption / reaction studies of organo-sulphur compounds and their load carrying mechanism, ASLE Trans., 16 (1973) 50.
45. T. Sakurai and K. Sato, Study of corrosivity and correlation between chemical reactivity and load carrying capacity of oils containing extreme pressure agents, ASLE Trans., 9 (1966) 77.
46. D. Godfrey, Chemical changes in the steel surfaces during extreme pressure lubrication, ASLE Trans., 5 (1962) 57.
47. R. O. Bjerk, Oxygen as an extreme pressure agent, ASLE Trans. 16 (1973) 97-106.
48. T. Sakai, T. Murakami and Y. Yamamoto, Optimum composition of sulphur and oxygen of surface film formed in sliding contact, Proc. JSLE Int. Trib. Conf. July 1985, Vol. 3, Tokyo, Elsevier.
49. M. Tomaru, S. Hironaka and T. Sakurai, Effect of some oxygen on the load carrying capacity of some additives, Wear 41 (1977) 117.
50. E. S. Forbes and H. B. Silver, The effect of chemical structure on the load carrying properties of organophosphorous compounds, J. Inst. Petroleum 56 (1970) 90.
51. H. E. Bieber, E. E. Klaus and E. J. Tewkbury, A study of tricresyl phosphate as an additive for boundary lubrication, ASLE Trans., 11 (1968) 155.
52. A. Dorinson, The additive action of some organic chlorides and sulfides in the Four-Ball lubricant test, ASLE Trans., 16 (1973) 22.
53. R. W. Mould, H. B. Silver and R. J. Syret, Investigations of the activity of cutting oil additives III. Oils containing both organo-chlorine and organo-sulfur compounds, Wear, 26, (1973) 27.

54. K. Kubo, Y. Shimakawa and M. Kibukawa, Study on the load carrying mechanism of sulphur-phosphorous type lubricants, Proc. JSLE Int. Trib. Conf. July 1985, Vol. 3, Tokyo, Elsevier.
55. H. A. Spikes, Additive-additive interaction in lubrication, Lub. Sci., 2 (1989) 3.
56. G. Heinicke, Tribochemistry, Verlag, Munich, 1984.
57. C. Kajdas, Importance of anionic reactive intermediates for lubricant component reactions with friction surfaces, Lub. Sci., 6 (1994) 203.
58. M. J. Furey, C. Kajdas, T. C. Ward and J. W. Hellgeth, Thermal and catalytic effects on tribopolymerisation as a mechanism of boundary lubrication, Wear, 136 (1990) 85.
59. P. A. Willermet, D. P. Dailey, R. O. Carter III, P. J. Schmitz and W. Zhu, Mechanism of formation of antiwear films from zinc dialkyldithiophosphates, Trib. Int., 28 (1995) 177.
60. P. A. Willermet, R. O. Carter III and E. N. Boulos, Lubricant derived tribochemical films- An infrared spectroscopic study, Trib. Int., 25 (1992) 371.
61. J. Schöfer, P. Rehbein, U. Stolz, D. Löhr and K.-H Zum Gahr, Formation of tribochemical films and white layers on self-mated bearing steel surfaces in boundary lubricated sliding contact, Wear, 248 (2001) 7.
62. S. Dizdar, Wear transition of a lubricated sliding steel contact as a function of surface texture anisotropy and formation of boundary layers, Wear, 237 (2000) 205.
63. S. S. V. Ramakumar, N. Aggarwal, A. Madhsudhana Rao, A. S. Sarpal, S. P. Srivastava and A. K. Bhatnagar, Studies on additive-additive interactions: Effects of dispersant and antioxidant additives on the synergistic combination of overbased sulphonate and ZDTP, Lub. Sci., 7 (1994) 25.
64. S. Pīaza, G. Celichowski and L. Margielewski, Load-carrying synergism of binary additive systems: dibenzyl disulphide and halogenated hydrocarbons, Trib. Int., 32 (1999) 315.
65. J. M. Martin, Th. Le Mogne, M. Boehm and C. Grossiord, Tribochemistry in the analytical UHV tribometer, Trib. Int. 32 (1999) 617.
66. C. Guo-Xu, X. Ren-Gen and D. Jun-Xiu, The tribological mechanism of action of oxovanadium (IV)- dithiphosphate, Lub. Sci., 9 (1997) 307.
67. A. Bhattacharya, T. Singh, A. P. Singh, R. Singh and V. K. Verma, Tribological studies of some N, S, and Cl containing extreme pressure additives, Lub. Sci., 7 (1994) 61.
68. W. Liu, Q. Xue, X. Zhang and H. Wang, Effect of molecular structure of organic borates on their friction and wear properties, Lub. Sci. 6 (1993) 41.
69. L. J. Bellamy, Infrared Spectroscopy of Complex Molecules, Chapman & Hall, 1975.
70. J. M. Hollas, Modern Spectroscopy, Wiley, New York, 1987.
71. J. Ferrante, Practical applications of surface analytic tools in tribology, ASLE Trans., 38 (1982) 223.

Nomenclature

A_r	real area of contact
c	concentration of the polar substance in the base fluid
E	heat of adsorption
f	overall coefficient of friction
f_m	friction coefficient for metallic junctions
f_l	friction coefficient for boundary film
F	friction force
k	rate constant in chemical reaction
K	equilibrium constant
N	carbon chain number
R	molar gas constant
s_m	shear strength of the metallic junctions
s_l	shear strength of the film
t	time
t_r	mean time of stay of a molecule at a surface site
t_x	time needed to move one molecular diameter
t_0	fundamental time of vibration of the molecule
T	absolute temperature
T_c	failure temperature
T_s	temperature of the surface
W	normal load
u	sliding velocity
x	diameter associated with an adsorbed molecule

Greek letters

α	fractional film defect
Δr	film thickness in EP reaction
ΔG^o	free energy change
ΔH^o	heat of adsorption
ΔS^o	entropy change
φ	fractional coverage of adsorbed molecules

5. Lubricated wear of metallic materials Theory and practice

5.1 Introduction

Lubricated wear occurs under mixed and boundary lubrication regimes and reduction of wear in these contacts has obvious economic advantage. The present chapter deals with boundary lubricated contacts only. Mixed lubrication, in which there is partial load support by EHD or hydrodynamic films will be considered in chapter 8 dealing with fatigue and wear. Lubricated contacts are far more common industrially and wear control is achieved normally through suitable additives. As discussed in chapters 2 and 4 wear controlling additives cannot be viewed in isolation from the rest of the additives in the formulation. Discussion with regard to wear here assumes that any formulation is already proven to be acceptable with regard to other performance requirements. The industrial needs in the area can be broadly categorised as follows:

1. Capability to predict at the laboratory level whether a new formulation results in lower wear in an existing system.
2. Ensuring performance when design changes are affected. Design changes usually involve more severe operating conditions.
3. Ensuring wear performance with new designs and materials for which available experience is inadequate. For example the needs of a ceramic bearing can be very different from a metallic bearing.
4. In all cases ensure reliability. From a lubrication point of view this means there should be no catastrophic transitions to high wear or scuffing.

The first section deals with lubricated wear in boundary contacts that are based on adsorption. The next section deals with the modelling of chemical wear. These models are viewed in terms of their practical applicability. Such a consideration identifies the research needs to strengthen the theory-practice interface. The third section briefly considers the problem of running-in. The empirical modelling of wear is considered in the next section with an example. The importance of such modelling has been brought out in the discussion. The problem of lubricant failure is addressed in the final section. The industrial needs broadly categorised above form the background for the present and subsequent chapters.

5.2 Wear modelling - adsorbed layers

5.2.1 Experimental variables

In this modelling only adsorbed boundary layers are involved. Models can only be validated by experiment and at the outset it is necessary to consider the role of variables in experimentation. As in the case of dry wear, wear rate refers to a quasi steady state situation. Any wear test goes through a stage of running-in until surfaces adjust to each other. Eventually a steady state is reached in the wear process. The running-in process is of industrial importance and will be dealt with later. Completion of the running-in stage even in laboratory machines is a function of the operating conditions and surface topography. In wear modelling it is necessary to ascertain that the running-in is complete and the wear rate does represent steady state.

The second issue in wear modelling is the estimation of contact temperatures. There is no common procedure in this regard. In some cases overall temperatures are considered while in other cases asperity temperatures are used. The methods adapted to calculate temperature rise again differ. An acceptable common procedure is needed at least for laboratory machines. Specialists in this area have to come together to evolve such a procedure.

The third issue is the possible interference of EHD or hydrodynamic effects. It is known that in many test configurations such effects can occur particularly at higher speeds. Wear may not be uniform over the surface and at microscopic level there can be convergent wedges leading to hydrodynamic action. This aspect is difficult to model and can lead to uncertainties in wear modelling.

The fourth issue is the nature of the test configuration selected. As pointed out in chapter 3 wear depends not only on the operating conditions but also on the test configuration. Care has to be exercised in generalising wear models obtained from one test configuration only. The various test configurations are considered in chapter 7.

5.2.2 Adsorption based models

The basic model available is that proposed by Rowe [1]. The model proposed by Stolarski [2] may be considered as a modification of this model. The problem treated first is for a single hydrocarbon lubricant. This system is less complex than an additive system where there is competitive adsorption between polar molecules and the hydrocarbon. First let the wear rate in the lubricated condition be formulated as follows:

$$\frac{V}{l} = K_b \left(\frac{W}{H} \right) \tag{5.1a}$$

where K_b is the boundary wear coefficient

Unlike dry wear K_b is a function of operating conditions. Rowe considered that for adhesive wear the wear depends on the extent of metal contact through the boundary film. Fractional film defect α can be expressed by Eq. 4.10 considered in the previous chapter. Thus in boundary lubrication actual metal contact area is expressed as αA_r. Since all asperity contacts do not wear a factor k_m was introduced. Rowe defined k_m as a true dimensionless constant specific to the rubbing surfaces and considered to be independent of any surface contaminant or lubricant. On this basis wear rate

$$\frac{V}{l} = k_m \alpha \left(\frac{W}{H} \right) \tag{5.1b}$$

The real area has been modified to consider the growth in contact area. This modification is derivable from Eqn.1.4 in chapter 1 which considered the growth in contact area. The final equation is

$$\frac{V}{l} = k_m \left(1 + 3f^2\right)^{0.5} \alpha \left(\frac{W}{H} \right) \tag{5.1c}$$

Expressing α as per Eq. 4.10 in the previous chapter

$$\frac{V}{l} = \frac{k_m x \gamma}{t_0 u} \frac{W}{H} \exp\left(\frac{-E}{RT_s} \right) \tag{5.2}$$

where $\gamma = \left(1 + 3f^2\right)^{0.5}$

The value of γ is close to unity since friction coefficient is usually less than 0.15 in lubricated contacts. The surface temperature rise was estimated with the low speed equation assuming heat flow to both the surfaces follows stationary source theory.

The temperature rise was estimated on the basis of the real area. The contact dimension was based on one equivalent circular area despite the fact that asperity spots are distributed over the worn area. This will overestimate the temperature rise at the asperities as can be seen from section 3.2.

The experiments were conducted in a pin-on-disk configuration with hemispherical copper pin sliding against a steel disk with n-hexadecane as the lubricant. The maximum load applied was 10 kg while maximum bulk temperature was 66°C. The maximum sliding speed used was 0.2 m/s. Taking $(k_m x/t_o H)$ as constant for a given system

$$\frac{V}{l} = C' \frac{W}{u} \gamma \exp\left(\frac{-E}{RT_s}\right)$$

where C' is a constant

Experiments were conducted to verify the above by varying one variable at a time and plotting the relationships. On the basis of slopes obtained E was estimated as 11,700 cal/mole, which is a reasonable value, based on the available literature. The correlations were inadequate when variations in wear rate were studied as functions of sliding velocity and temperature separately. The variation with load showed better correlation. The intercept obtained when wear rate (V/l) is plotted against $(\gamma / u)\exp(-11{,}700/RT_c)$ on log-log plot was used to estimate C'. 'x' was obtained from the available literature. Determination of t_o from fundamental considerations presented problems and two possible t_o values of $2.0\text{x}10^{-13}$ and $2.8\text{x}10^{-12}$ were estimated from the available theoretical approaches. The latter value was justified on the grounds that it leads to a k_m value of 0.23. This is close to the value of 1/3 in Archard's wear equation (3.15d) in which all junctions lead to wear. This amounts to an assumption that *every* adhesive junction leads to a wear particle.

Stolarski [2] proposed a more detailed model for lubricated wear. The author considered both partial EHD and boundary conditions. He used a different approach to the boundary lubricated wear and separated the real area (A_r) into plastic (A_p) and elastic (A_e) areas giving

$$A_r = A_p + A_e$$

It was considered that adhesive wear should be based on A_p. The model was tested with ethylene glycol and n-hexadecane as lubricants in a pin-on-disk machine. The rotating low carbon steel disk was slid against a hemispherical brass pin. The variables were tested one at a time. The reported correlation between theory and experiment was good. The author stated the limit of applicability of the equations was up to a sliding speed of 0.1 m/s for which no reason was mentioned. The fractional film defect α was obtained by the same procedure as used by Rowe but using the double exponential expression without simplification. It may be mentioned that one series of experiments conducted by Stolarski [3] showed that when sliding speed was changed over a wide range for a fixed sliding distance the experimental wear volume was nearly constant. The experiments clearly fit Archard's law with a constant wear coefficient. One of his figures illustrating the speed influence on wear for brass-steel combination using hexadecane is given in Fig. 5.1. The validation of the model is inadequate.

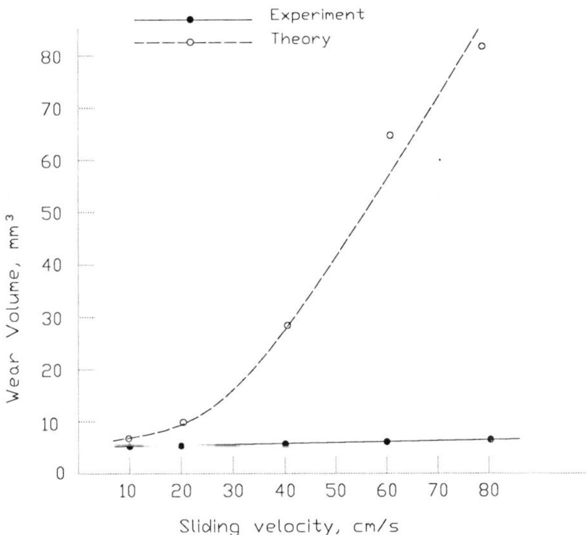

Fig. 5.1. Wear volume Vs sliding distance for a constant sliding distance of 1000 m at 50 N load. From Ref [3].

Mention may be made of another model for boundary lubrication developed at IBM mainly for business machines and described by Bayer [4]. The approach is based on the concept of zero wear. The number of cycles the system can survive with zero wear is a function of the maximum shear stress in the contact and the nature of the lubricant. Zero wear is defined as a situation in which the wear is confined to the asperity dimension. The model had been successful for the lightly loaded slow moving contacts involved in such machines. It is difficult to extend the model to normal wear situations.

5.2.3 Assessment of the adhesive wear model

A detailed fundamental analysis of the model due to Rowe was carried out by Beerbower [5]. He considered that the heat of adsorption for steel and copper will be different. Another interesting point raised by him is the probability of contact. The film defect considered by Rowe was with reference to one surface. If the film defects for both the surfaces are α_1 and α_2 then the probability of contact will be $\alpha_1\alpha_2$ which will be much lower than either of the individual values. In chapter 4 the problems involved in assigning a specific value to E have been discussed. It can vary significantly depending on the nature of oxides and metals. Even with pure hydrocarbons oxidised polar compounds can form and this can lead to shifts in the heat of adsorption. The present author considers E to be a variable particularly for real systems.

Another issue to be considered is the modification done to basic Archard's equation. The equation as expressed (and ignoring γ) takes wear rate as proportional to real metal contact area, (αA_r). No consideration is given to the size of the transfer particle. As discussed in section 3.4 the adhesive wear particles through oxide films will have on average a diameter of ($\alpha^{0.5}d$) where d is the asperity diameter and α is the film defect. It is to be emphasised that this smaller wear particle will result from a sliding distance of d. The idea can be now applied to a boundary lubricant by replacing the oxide film with a boundary film with the same conclusion. There is no oxide now and all the contacts at the defects are metal to metal. With this concept the wear rate will be

$$\frac{V}{l} = k_m \alpha^{1.5} \left(\frac{W}{H} \right)$$

The value of 1/3 in Eq. (3.17) is not used here as k_m in the above equation approximates this value. On this basis for a given wear rate the α values will be much higher. As an example in a test at 8 kg load [1] the wear rate was $2\text{x}10^{-8}$ cm^3/cm. The value of α as per Rowe's equation will be $6.7\text{x}10^{-5}$ for this case. From the above equation α will be $1.6\text{x}10^{-3}$. This approach is a suggestion based on physical reasoning. The particles in real situation will have a distribution. So consideration on the basis of α alone will be approximate. While the suggestion here is made regarding boundary lubrication the influence of the size factor is general and should be considered whenever removal is on a scale lower than the asperity contact dimension. One additional aspect is the assumption that a hemispherical particle is involved in dry and lubricated contacts. This is a convenient way to assess size effects. Even if particles are not hemispherical the size effect persists though the equation will be modified.

Plastic contact at asperities will be significant for soft materials like copper and brass used in these tests. It is logical to expect adhesive wear to be more likely with plastic contacts and so the separation of asperity junctions into plastic and elastic contacts by Stolarski appears logical. But even in plastic contacts the oxide must affect the metal to metal contact. This influence was not taken into account by Rowe or Stolarski. At the other extreme, suppose all contacts are elastic and there is no metal contact through the oxide films. In such a case the wear will be governed by the oxide removal through a fatigue process. The oxides are usually thin in lubricated contacts and the type of equations used for oxidative wear as discussed in chapter 3 are not applicable. Lubricant affects the wear process by reducing the extent of oxide contact. Lubricants also reduce frictional traction that increases the fatigue life. For the intermediate situation the overall wear will be governed by the extent of metal contact in all the junctions. In case the metal contact levels are very low the oxide removal rate is expected to govern the wear rate.

Another question is the actual mechanism operating within the contact zone. The adsorption model assumes that vacant sites on the surfaces lead to metal/oxide contacts. Another possibility is to consider that the contacts arise due to local rupturing of the shearing molecular layer. The film itself has defects, which in turn influence the local contact. The contact through boundary layers is expected to be a function of the shear rate and temperature within the film. The modelling in this case depends on the complex rheological behaviour of the film, which cannot be predicted at this stage. The rheological behaviour of monolayers and multi-layers in nano contacts is being modelled at very low speeds and loads as mentioned in the previous chapter. Any extension to high shear rates will be useful for practice. This aspect is also of importance with regard to lubricant failure as will be considered later in the chapter.

Another related problem is the adsorption/desorption phenomenon in the high pressure contact zone. Unlike the open system where the molecules can adsorb or desorb from or to the bulk fluid, the monolayers in contact are highly confined. Adsorption and desorption can occur between the two asperities in which case the probability of defects will be reduced.

A more complex relation can be set up when a polar additive is added to the hydrocarbon base fluid as reported by Rowe [6]. The surface is now covered with the molecules of base fluid as well as additive. The additive concentration on the overall surface, θ, is obtained by a relation based on the Langmuir relation given by Eq. 4.2 in chapter 4. The surface defects for base fluid and the additive covered zones can be obtained separately on the basis of this equation. It is then assumed equilibrium establishes itself in the contact zone and the equilibrium constant is related to the change in heats of adsorption of the base fluid and the additive, and the overall change in entropy of the system. Only the final equation for the case where molecular size of base fluid and additive is the same is given below.

$$\left(\frac{V}{l}\right)_c' = \frac{\exp(-\Delta E/RT_s)}{\exp(\Delta S^0/R)}\left[\frac{\left(\frac{V}{l}\right)_b' - \left(\frac{V}{l}\right)_c'}{C}\right] + \left(\frac{V}{l}\right)_b' \frac{\exp(-\Delta E/RT_s)}{t_0'} \quad (5.3)$$

where

C = additive concentration, mole fraction

$\left(\frac{V}{l}\right)'$ = wear rate x $\sqrt{1+3f^2}$ and subscripts b and c refer to the base fluid and additive containing oil at concentration C

ΔE = *difference* in heat of adsorption of additive and base fluid

ΔS^0 = Overall entropy change

t_0' = ratio of fundamental time of vibration of additive and base fluid molecules

The SI units with regard to the above parameters are given in chapter 4. Rowe used CGS units in his work. Linear plot between $(V/l)_c'$ and the square bracketed portion can be drawn and from the slope and intercept ΔE and ΔS° can be obtained indirectly. Limited testing of the model was done by finding wear rates at different concentrations of octadecanol, stearic acid, and n-octadecylamine in hexadecane. It is difficult to comment on this model except to say that it will be subject to the same problems as in the earlier case. The authors found that wear rates decreased substantially with additives. Also for a given additive the wear rate decreased with increasing concentration. As discussed in chapter 4 lubricant failure temperature increases with the additive concentration as well as the heat of adsorption. These findings taken together show higher concentrations of the additive are advantageous from the wear and scuffing point of view as expected.

The influence of test configuration has already been discussed at the outset of the section. These aspects are general and are relevant to testing with antiwear additives as well. Any evaluation of scuffing properties is also influenced by the test configuration.

5.2.4 Application of the model

Effective quantitative use of the model is not possible due to the uncertainties discussed above. The model is inadequate even with a simple hydrocarbon system in a limited operating range. The model was considered in some detail to appreciate how fundamental approach was attempted. The interaction of variables on the wear behaviour is complex and cannot be modelled on the basis of physical chemistry considerations alone. The way out is to develop empirical relations as will be considered later. Such relations suffer from their restrictive nature and should be utilised to develop insight into the mechanisms involved. This in turn can lead to better scientific modelling. Qualitative considerations based on the above model are useful. The influence of heat of adsorption based on this model, as well as earlier observations, is clear. The obvious choice is to use additives with higher heats of adsorption wherever possible. In the rubbing contact the value of heat of adsorption is variable due to modifications in the surface and the lubricant. But it can be argued that it is an advantage to start with an additive having higher heat of adsorption. The large variation in wear rate with sliding velocity predicted by theory is doubtful. The increased wear rate with temperature is generally observed but it is unlikely to be exponential.

The nature of fats and oils used for industrial purposes is varied and they cannot be defined like pure compounds in fundamental studies. Their use is increasing because of environmental reasons as discussed in chapter 2. There is already a large amount of testing activity on potential materials for several applications that include hydraulic fluids and gear oils. At the other end there is increased use of synthetic esters due to their superior performance in comparison to mineral oils. Some of the esters also have better biodegradability, which is an additional advantage. Most of the testing activity is empirical because formal models discussed above are inadequate.

Several fat based materials are conventionally used in metal working operations mainly as stable and metastable emulsions. In metal cutting, wear of the work piece is to be maximised while at the same time minimising the tool wear and pick-up. In metal forming like rolling roll wear and pick-up is to be minimised while the reduction per pass is to be maximised. Another important requirement is the surface integrity of the work in terms of brightness, surface roughness, and freedom from stains. In many of these operations the temperatures are significantly higher in the contact zone despite water cooling and can easily exceed the melting points of the soaps. It may be mentioned that some fatty materials and synthetics have been successful even in the hot rolling of steel. Conventional ideas can only predict lubrication failure in such cases. This area will be considered in chapter 7 but will be confined to performance evaluation.

5.3 Chemical wear

Chemical wear may be defined as the wear observed with chemical additives that are intended to react chemically with the surface. In chapter 4 the mechanisms of action of additives were considered. While these studies provide insight into the overall behaviour, wear modelling is only possible through an understanding of the rate of formation and the removal process. To focus attention, only steady state wear is considered assuming boundary lubrication conditions. The issues related to scuffing and protection with EP additives will be considered in the final section. The next part deals with the wear mechanisms. The possibility of modelling with formal equations is limited and is considered where appropriate.

5.3.1 Wear mechanisms and modelling

Consider two asperities coming into contact. Several layers of molecules consisting of the additive and base fluid get dragged into the contact and it is assumed that by the time the maximum contact pressure is reached monolayer surface coverage of the additive remains on both surfaces. As the contact moves two competing mechanisms operate. One is the rubbing process resulting in wear, while the other is the chemical reaction of the adsorbed additive at the higher temperature and stress conditions in the asperity. Chemical interaction can be thermally activated. It can also be due to exoelectron emission. If the layers provide adequate protection against wear the asperities separate with some degree of reaction on their surfaces. Oxidation can occur due to dissolved oxygen diffusing to the surface and competing with the additive molecules for surface reaction. As the asperities move out of contact the temperature decreases. Additional reaction is now possible due to the easy access of the additive molecules and oxygen as the asperity cools. Depending on the bulk temperature there can be further reaction at the surface. This reaction will continue till the asperity again gets engaged in the contact zone. The chemical interaction zones are illustrated in Fig. 5.2. The surfaces are covered with the lubricant. The reactions occur at the asperity level and at the bulk temperature on the wear track. The processes repeat with each cycle resulting in the growth of film thickness. As per conventional approach the film detaches when a critical thickness is reached.

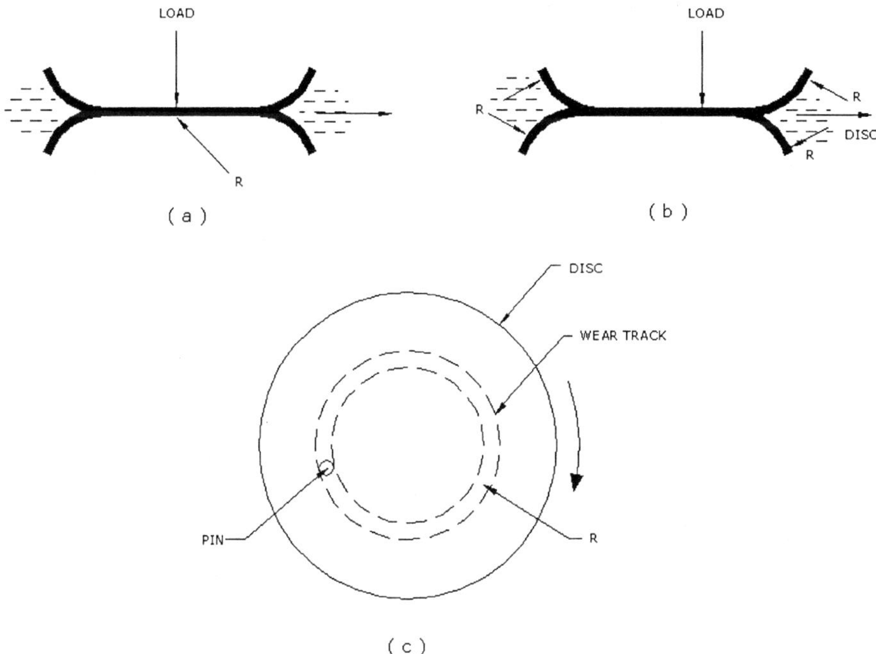

Fig. 5.2. Chemical reaction zones represented by R. (a) Reaction at high asperity temperature (b) Reaction in the asperity cooling zone and (c) Bulk reaction on the wear track.

The generally accepted view of chemical wear [7] may be represented schematically as shown in Fig. 5.3. The wear rate is plotted as a function of temperature for two additives for a given load and speed condition. Chemical wear, governed by the reaction films is expected to occur when the film thickness is adequate, leading to wear within the film itself. In this zone, to the right of minima, the wear is governed by the chemical reaction rate. Higher reactivity leads to increased chemical reaction resulting in increased chemical wear. At any given reactivity the removal and formation rates are balanced and a particular reaction rate is established. Under conditions where the film formation is inadequate, adhesive wear is considered to occur in portions of the contact leading to higher wear. This zone is to the left of the minima. In this zone the wear decreases with temperature as less metal contact occurs with increasing film formation. At an optimum temperature the minimum thickness is just enough to sustain chemical wear. At this point the wear is minimum. Different additives will have different relationships as shown for additives A and B.

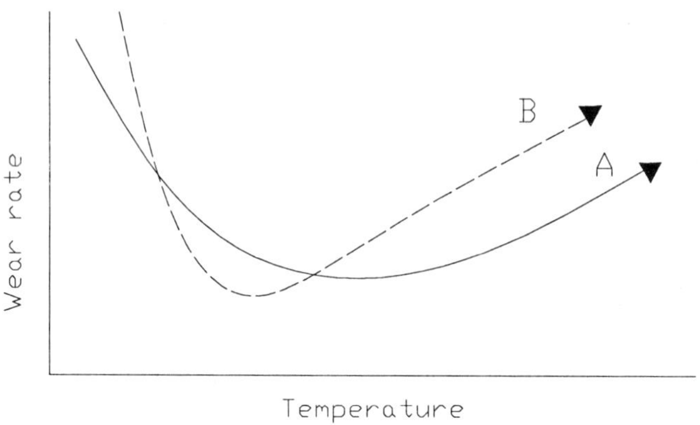

Fig. 5.3. Wear rate Vs temperature for additives A and B.

5.3.2 Limitations of the available model

One possible way to model the chemical wear process is to invoke the oxidative wear model of Quinn discussed in chapter 3. The concept can be applied to the wear of chemically reacted films. Film wear modelling has two components. One is the simple expression based on wear coefficient and the other is the estimation of wear coefficient based on kinetic treatment. These were discussed in section 3.7.2. For clarity in this sub-section the film wear model refers to the first component while the kinetic model refers to the second component. In the present case oxidation and additive reaction occur together at the surface. The chemical reaction will be a function of the additive concentration. The oxygen availability will be a function of oxygen solubility. Also oxidation reaction will be a function of additive concentration since oxygen competes with the additive for surface reaction. For a given additive concentration it is assumed that the overall film growth is governed by a parabolic law. This approach is similar to that used for oxidative wear modelling. Following the approach given in section 3.7.2 and assuming steady wear, the film wear rate may be expressed by Eq. (3.20b) as follows:

$$\frac{V}{l} = K_f\left(\frac{\xi}{d}\right)\left(\frac{W}{H}\right)$$

It may be recalled that K_f is the probability of wear and is equal to the inverse of the number of cycles needed to form a film of given thickness.

Since many real situations involve elastic rather than plastic contacts H needs to be replaced by the average elastic stress at the asperities. For the present argument H is retained as the flow pressure. Rowe [7] has tabulated wear coefficients for several chemical additives based on 4-Ball tests. The overall wear coefficient K' obtained at 50°C refers to the value as obtained from

$$\frac{V}{l} = K'\left(\frac{W}{H}\right)$$ where K' is equal to $K_f(\xi/d)$

The K' values for good antiwear agents range from $0.1\text{-}10.0 \times 10^{-8}$. Consider for the present purpose a typical value of 10^{-8}. Considering an asperity contact diameter of 10 μm and a critical film thickness of 1.0 μm the value of K_f will be 10^{-7}. In other words on average the film thickness grows to 1.0 μm in 10^7 cycles. If the parabolic law is recast in terms of cycles

$$\Delta r^2 = k_n n \tag{5.4}$$

where n is the number of cycles, Δr is film thickness, and k_n is a dimensional constant

It can be shown that per cycle growth at $n = 10^7$ will be as low as 5×10^{-8} μm. This growth per cycle amounts to 5×10^{-4} Å. Even at n = 1000 the growth rate per cycle amounts to 5×10^{-2} Å. Even if a lower critical removal thickness of 0.1 μm and higher K_f of 10^{-6} are assumed, the growth rate per cycle amounts to a value less than an angstrom. These growth rates are physically inconceivable and it is not possible to apply kinetic model based on parabolic law. The mechanism involved has to be treated differently.

The above argument is with regard to additive systems used in practice and which provide low wear. Nakayama et al [8] modelled wear with a copper-steel system using elementary sulphur as the additive. The pin-on-disk machine was operated with sulphur concentrations ranging from 0.01 to 0.05% by weight and the wear of the stationary copper pin was monitored. Their results followed the behaviour schematically shown in Fig. 5.3. The film wear to the right of minima was modelled kinetically as per the oxidative wear model. It may be recalled that for a constant critical film thickness $\ln(V/l)$ is proportional to $(-Q_p/RT_s)$. When $\ln(V/l)$ is plotted against $1/T_c$ the value of Q_p can be estimated from the slope of the line. This is a common procedure to obtain the activation energy. While reasonable relations were observed by the authors for each concentration, Q_p values had a wide range of 8.9 to 18.2 Kcal/mole. Since Q_p is expected to be constant from the postulated model such large variations are a problem especially because wear rate is exponentially related to Q_p. Similar problems exist with oxidative wear. Problems like possible variations of critical film thickness, variation in activation energy and Arrhenius constant, and the accuracy of the temperature estimates by existing theories is not discussed further. One parameter affects the other and the problem is difficult to resolve. But one question that arises is the extent to which such a model can be used. The values of K_f calculated from the available information in this paper range 10^{-2}-10^{-4} if a critical film thickness of 1.0 μm is assumed. Such high wear rates will be unacceptable for lubricated contacts. This example is cited to show that in the case of high film wear approximate modelling is possible. The concepts involved in the model have relevance for situations involving EP action. In EP action the possibility of scuffing is controlled by fast reaction with active additives.

5.3.3 Alternative proposal for wear mechanism

As discussed above the wear mechanism cannot be reconciled with the growth and removal of critical film thickness. One possibility for low K_f in lubricated systems is the influence of adsorption on the reacted film. This adsorbed layer consisting of additive and base fluid molecules will reduce contact over the film area just as in the case of boundary lubrication with adsorbed layers. In such a case the film contact area is reduced and the wear will be lower. On the other hand many antiwear additives are known to function effectively above 100°C. Adsorption influences at such temperatures are expected to be lower. In some cases there can be strong adhering films deposited over the reacted film. The reacted film here refers to the metal additive reaction. The deposited films can arise from polymerisation or other reactions within the lubricant system. The other aspect is the growth of reaction film. It is possible that progressive reaction and slow building up develops a uniform film with very few defects and pores. Such a film offers a very strong barrier to reaction restricting film thickness to very low values.

Removal of such films is likely to be governed by fatigue. Fatigue cycles for removal will be related to the normal and tangential stresses. Defect free thin films are expected to survive a large number of cycles. In thicker films cracks can propagate from the defects leading to earlier failure. Chemical reaction will be restricted to the small failed zones and may be considered as a repair action that re-establishes the uniform film. The role of adhering boundary layers is an additional variable which has to be taken into account. But eventually the reaction films wear and in so far as their wear is concerned, their mechanical response to cyclic stressing is of importance.

5.3.4 Assessment of the proposed mechanism

The first issue is whether an assumption can be made that a strong barrier to reaction can occur with virtually no further growth of the film. As stated by Molgaard [9] the parabolic oxidation law is applicable when film thickness is higher than 0.1 μm. At lower thickness the oxidation mechanisms are far more complex. One possibility is that they follow a logarithmic law [10] as expressed below:

$$\Delta r = k_e \log(a_1 t + 1) \tag{5.5}$$

where k_e is a rate constant and a_1 is a constant.

Such an equation predicts a much lower growth rate for the film as a function of time. Assuming such relations are possible with the additive system the negligible growth rate of the film is directionally explicable.

A second issue is the evidence with regard to film thickness. Investigation reported at the laboratory level with zinc dithiophosphate by Vipper et al [11] is of interest in this regard. They studied the influence of copper naphthanate concentration on the antiwear and EP action of dithiophosphate and observed that an increased percentage of the naphthanate improved the wear performance. The antiwear films were analysed and the depth of the chemically reacted film referred to as SS by the authors was inversely related to the wear. Thus the thinner the reacted film, the better the wear reducing capability of the film. At the minimum wear rates observed the films involved were of the order of 50 nm while at the higher wear rates the films were in the range of 100-300 nm. The SS layer is the zone that involved diffusion of oxygen and sulphur. They explained the mechanism as being due to the formation of a complex with the copper compound and reduction of surface oxidation resulting in thinner films. The authors also took into account the physically decomposed layer (PD) and the hardened zone due to carbon diffusion called B layer and considered the SS layer to be of primary importance. The wear rate at the low SS levels was about 0.1x10^{-3} mm^3/km while at the high SS values the wear rates ranged from 0.5 to 0.7x10^{-3} mm^3/km. Reacted films on engine liner surfaces analysed by Becker and Ludema [12] also showed reacted films of around 0.1 μm. This film is attributable to the action of zinc dithiophosphte additive and oxygen in the engine oil. From this evidence it is reasonable to postulate that very thin films are involved in good antiwear action. While oxygen is involved the nature of oxides and their role is not well studied in antiwear films. Ludema [13] considers that Fe_3O_4 is the desirable oxide at least in the ring-liner tribology. A typical example of low SS thickness with copper napthanate at 0.05% copper concentration and dithiophosphate additive DF-11 is illustrated in Fig. 5.4. The SS layer is about 20 nm. This diagram is based on depth profiling based on AES analysis. The thickness is based on the sputtering rate and the time involved to reduce the concentration of active elements as shown in the figure.

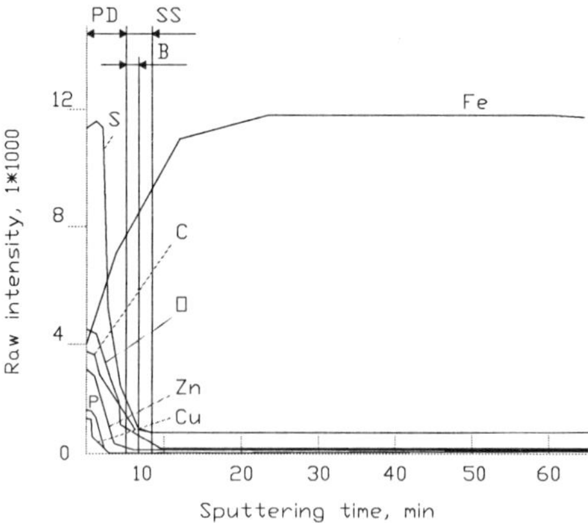

Fig. 5.4. Depth profile of reaction film with zinc dithiophosphate. (Reproduced from Ref. [11] by permission of Leaf Coppin Publishing Ltd.).

The above assessment is satisfying with regard to a workable mechanism for antiwear additives. At least with one class of important additives there is support for the postulate that thin films offering strong barrier to reaction provide good antiwear action. But it is known that glassy deposited films are involved with dithiophosphates. As their role in the wear process is not clarified, it amounts to an assumption that overall wear rate is influenced by the extent of sulphur and oxygen diffusion into the metal. The argument may be extended to the general class of antiwear additives many of which do not form deposited films. The equilibrium film thickness will be related to the nature of additives involved in a given system. But it is expected that for a given system the film thickness that offers a maximum barrier to reaction corresponds to best antiwear action. The proposed approach is apparently similar to the conventional film wear concept shown in Fig. 5.3. But the important difference is that in the proposed mechanism wear is not governed by growth and removal of a film with critical thickness. It is governed by cyclic stressing and removal of the film which is a fatigue process. On this basis the fatigue removal of the reaction film is the rate determining step. The role of chemical reaction is one of re-establishing the thin reaction film at the worn zones. The overall film thickness involved is very low and is usually less than 0.1 μm. The response of the system to the operating conditions will be different and is no longer predictable on the basis of the conventional model. If there are thin deposited films they may offer a further barrier to reaction.

5.3.5 Fundamental considerations

The proposed mechanism is logical but based on limited data from literature. Fundamental investigations are necessary to establish the mechanism. One key issue is the barrier nature of the films. Direct evidence with regard to this aspect should be sought. The other aspect of importance is the study of wear particles. These issues are discussed below.

5.3.5.1 Nature of barrier films

It is necessary to develop direct evidence of the barrier nature of a film on the wear scar. This will be referred to as post wear film. An example of such a study in EP lubrication by Sethuramiah et al [14] may be cited. When tests were conducted by step-load procedure in a 4-Ball machine with diphenyl disulphide (DPDS) and elementary sulphur the failure load was higher for DPDS in comparison to sulphur. The detailed methodology is given in the cited reference. Normal testing involving fresh sets of steel balls at each load gives a higher failure load for sulphur and is attributed to its higher reactivity. It was postulated that in step-load tests the lowering of surface roughness with the less reactive additive provided partial hydrodynamic load support. Such smooth surfaces can be a result of lower reaction. The point of interest here is a comparison of the scars obtained in step-load tests with both the additives under similar conditions. This unpublished work is cited here. The sulphur content on the wear scars obtained at 100 kg was first assessed by EPMA. The surfaces were then reacted in static conditions under argon atmosphere with 0.294% sulphur solution under identical conditions for 30 minutes at 180°C. The increase in sulphur content with the sulphur scar was 87% in comparison to 26% for DPDS scar. It is of interest here to note that when the reaction is conducted in normal air atmosphere the sulphur content decreased by 22% for the DPDS scar while it increased by 51% for the sulphur scar. This is an example of competitive reaction of sulphur and oxygen depending on the nature of surface films. This means films formed with DPDS offer a stronger barrier to reaction. Better techniques like AES can now be used to study the reactivity of post wear films more precisely. Modern electrochemical techniques may also be useful for this purpose. Thus direct investigation of the barrier nature of the films is feasible and is worth pursuing. Such studies can be more easily conducted on laboratory specimens, but some reference surfaces from real systems will be useful for comparison.

The mechanical characterisation of the post wear films is hardly attempted so far. Extensive research is being conducted on the adherence and durability of coatings by scratch tests and other techniques. Adaptation of these techniques to study antiwear films will be valuable. Such characterisation will form a useful link between the chemical and mechanical nature of the films.

5.3.5.2 Analysis of wear particles

A wear particle is directly related to the wear process. Wear particles are analysed for their morphology and size range in oil condition monitoring. Oil condition monitoring refers to the assessment of equipment condition by periodically examining wear debris. Such an analysis detects the onset of malfunctioning by observing the changes in particle size distribution and their nature [15]. Special techniques like ferrography [16] are used to separate the particles size wise. Elemental analysis supplements the study of particles. For example a sudden increase in copper content can mean impending failure of the copper based bearings. Detailed analysis of the wear particles for their structure and composition is what is of importance in wear mechanisms unlike the global composition studies conducted in condition monitoring. Such studies are conducted in great detail on worn surfaces but there are only few studies reported on wear particles in lubricated contacts. On the other hand detailed analysis of wear particles is quite extensively reported for dry wear. One interesting study [17] pertains to the study of wear fragments obtained with zinc dithiophosphate additive in a steel-cast iron system. The flakes consisted of amorphous regions in which very small crystalline iron sulphide particles were observed. The amorphous region contained zinc, iron, phosphorous, and oxygen but no detectable sulphur. The amorphous region was likely to be glassy phosphate containing iron and zinc. Such studies open up interesting possibilities. For example is the iron sulphide obtained due to partial contact through glassy films? Does the absence of oxygen in crystalline material support the current hypothesis [18] that iron oxides are 'digested' in developing glassy films? Such broadening of knowledge is possible by a detailed analysis of wear particles. Basic investigations of this kind coupled with surface analysis will clarify wear mechanisms much better. The available detailed information on surface films is not effectively linked to the removal process involved. These links can be established through wear particle analysis. The problems involved in developing this kind of technology should not be underestimated. Wear particles may be clustered, structures may not be uniform, and wear particles may be mixed from the initial running-in to later film wear. The analytical tools needed for detailed study may not be generally available and detailed analysis can be undertaken only by specialists. With the available technologies in condition monitoring, including on-line monitoring, it should be feasible to provide representative samples for a given condition. Also, as is the situation in any technology, there will be limits within which one has to work. Even with this scenario a lot of progress that is relevant should be feasible. If such attempts are not made trial and error approaches that are time consuming and limited in scope have to continue.

5.3.6 Practical aspects

The proposed mechanism lays emphasis on the characterisation of post wear films and wear particles. Besides helping in eventual modelling such characterisation is of direct practical relevance. Comparison of films in the laboratory tests and real systems will be useful in establishing the operating conditions in a machine that correlates with practice. In many cases it is impractical to simulate the real systems in the laboratory. This is because there are a large number of tribological systems operating over a wide range of operating conditions. To make a laboratory rig more widely applicable the zones of applicability to real systems should be identified. This identification is possible through the analysis of post wear films and wear particles as suggested above. The idea is to simulate the wear process instead of the test rig. Becker and Ludema [12] studied ring-liner tribolgy in a laboratory reciprocating test. It is of interest here to note that the validity of the simulation test was based on a comparison of wear particles and surfaces obtained in engine tests and the simulator. Morphology was the main aspect considered. As stated by them the operating conditions of an engine are not simulated and they relied on the simulation of the wear process. Detailed analysis of wear particles and post wear films as suggested above will strengthen such simulations.

The complexity of chemical reactions is not unique to tribology. Many chemical reactions in real systems cannot be modelled with laboratory information alone. Catalytic reactions, for example, in hydrocracking are very complex. Industrial realisation of such processes involves different levels of empiricism. Solutions emerge because of the large scale R&D effort at different scales. This effort is justified because of the significant economic benefits accruing from such technologies. The suggestions made to broaden the understanding of the additive action are worth pursuing. The extent to which investigations can be conducted depends on economic considerations. To start with, investigations with engine oils can be economically justified. These oils are consumed in large quantities and improvement of engine life is an important consideration.

5.4 Running-in

Running-in refers to the adjustment of newly assembled components under controlled conditions. Freshly assembled surfaces are not well matched and can have misalignment at the micro level. This results in non-uniform loading with some zones highly stressed in comparison to others. If the design load is applied right in the beginning there can be large plastic deformation at some patches. Such patches can result in large scale adhesion leading to scuffing and failure of the components. To avoid this situation the main strategy employed is to run the system at moderate operating conditions to start with, and progressively adjust the surfaces to each other. The severity on the components is gradually increased to the design levels. If the running-in is incomplete there is a possibility of failure in service. While well run-in components do not fail there is always a possibility of a small percentage of failures. This is because of the statistical nature of micro level misalignments with some odd components unable to run-in effectively with the prescribed procedure.

The procedures adopted for running-in are empirical and based on experience. Running-in is of importance in bearings, gears, engines, and other components. Major interest is in the ring-liner contact of engines where contact conditions are severe. Criteria to assess the completion of the running-in are usually based on the final expected roughness reached. The characterisation of roughness itself can have varied levels of sophistication, particularly with reference to engine liners. Some limits also apply with regard to the quality of the surface based on microscopic observation. These criteria as suggested in some research papers [19,20] as well as manuals of engine manufacturers are empirical and based on experimental observations. One different and interesting approach reported by Joseph and Raman [21] is based on determining the variations in compression pressure during operation. The authors showed that maximum compression pressure is reached when running-in is complete. The pressures were measured by a piezo electric sensor. The running-in done in the industry basically completes the initial and difficult phase. The completion of running-in is gradual and may take easily more than hundred hours in an engine.

The industry is interested in reducing the time needed for running-in. They are also interested in well defined criteria to decide the completion of running-in. Yet another need is the procedure for running-in. Empirical methods have their problems. When a design is changed and the operating conditions are more severe the criteria have to be re-established by trial and error, which is a time consuming process. This also applies to situations where surface coatings or materials are changed. The available answers from tribologists are limited in the area. Some partial answers will be attempted in the relevant parts of the text.

The general nature of the running-in process followed by a steady state is illustrated in Fig. 5.5. The overall process may be described by the following equation [22]

$$\dot{V} = \left(\dot{V}_0 - \dot{V}_s\right)e^{-bt} + \dot{V}_s \tag{5.6}$$

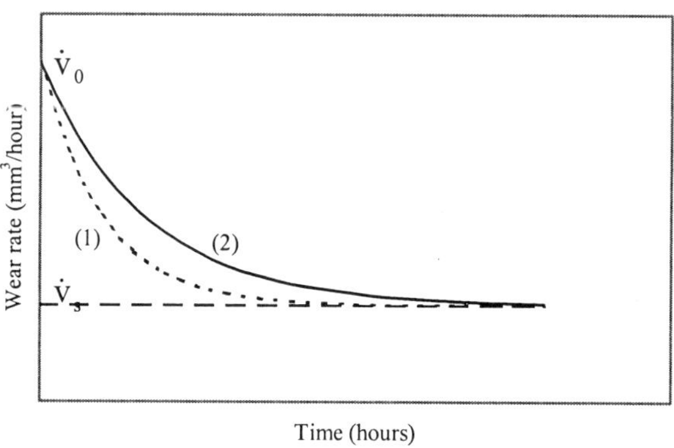

Fig. 5.5. General nature of running-in behaviour. The two curves represent two different 'b' values given in the equation.

where $\dot{V}, \dot{V}_0, \dot{V}_s$ represent wear rate at time t, wear rate at t = 0, and steady state wear rate respectively. The rate is expressed as wear volume per unit time.

The two curves represent two different values of the exponents b involved. The running-in in these curves is satisfactory in the sense that the process gradually changes to steady state. If the running-in stage is ineffective there can be large fluctuations in this zone with the associated risk of scuffing. One theoretical model is available which looks at the process only in terms of change in roughness [23]. As running-in is a complex process with several processes of adjustment, it is better to concentrate on the overall model as expressed by the above equation.

5.4.1 Modelling running-in and steady state wear

The complexity of wear mechanisms has been discussed earlier. Empirical modelling is a practical way to study the influence of various parameters on the wear rate. It is firstly necessary to quantify the running-in and steady state wear precisely. Once this procedure is available the influence of operating variables can be studied systematically by designed experiments. Rajesh Kumar et al [24] have recently reported such a methodology. The concepts and the final results are discussed here. Numerical procedures used are given in detail in the reference and not given here. The procedure starts with the reasonable assumption that the running-in and steady state wear follow the exponential relation given in Eq. 5.6. Integration of this equation with the boundary condition $\dot{V} = \dot{V}_s$ at $t = \infty$ and $V = 0$ at $t = 0$ we obtain

$$V = a\left(1 - e^{-bt}\right) + \dot{V}_s t \tag{5.7}$$

where

$$a = \frac{\left(\dot{V}_0 - \dot{V}_s\right)}{b}$$ and V represents wear volume at time t

In the selected reciprocating test wear volume was obtained as a function of time. The value of b was obtained by an iterative procedure such that $|1 - R|$ is less than 1×10^{-4} where R is the coefficient of determination.

The methodology leads to a proper statistically based determination of the running-in and steady state wear rates.

The experiments were conducted in a reciprocating tester with an EN 31 steel ball sliding against a flat of the same steel. Tests were conducted at a constant frequency of 50 Hz and a stroke of 1.0 mm. Commercial engine oil with a viscosity of 129.9 cSt at 40°C and 13.3 cSt at 100°C was used. The oil contains zinc dithiophosphate as antiwear additive and the zinc and phosphorous contents were 742 and 1890 ppm respectively. Sulphur content was not determined. The variables selected were

Load 20N, 40N, 60N
Roughness (R_q) 0.35 μm, 0.55μm, 0.75 μm
Bulk temperature 50°C, 100°C, 150°C

The roughness refers to that of the disk. The experimental design consisted of one-third fraction of 3^3 factorial design. Each test was of eight hours duration with ten steps. At each step the wear scar on the ball was measured from which the wear volume was obtained with known equation that takes elastic recovery into account. The disk wear was determined after the final stage by measuring the scar dimensions. The surface temperature rise was estimated on the basis of geometric area only and was added to the bulk temperature to obtain the contact temperature. Estimation on the basis of asperity temperatures was not done in this case.

Firstly data from each experiment were fitted to a regressed curve as discussed earlier. The nature of fit observed is given for one case in Fig. 5.6. From this equation the necessary wear parameters were calculated. The next step was to obtain the empirical equation that fitted all the experiments with suitable exponents on the variables by statistical methods. The final equation obtained for steady state wear of the ball is given below:

$$w_s = 1.37 \times 10^{-4} P^{0.74} R_q^{0.98} T_c^{-0.79} \tag{5.8}$$

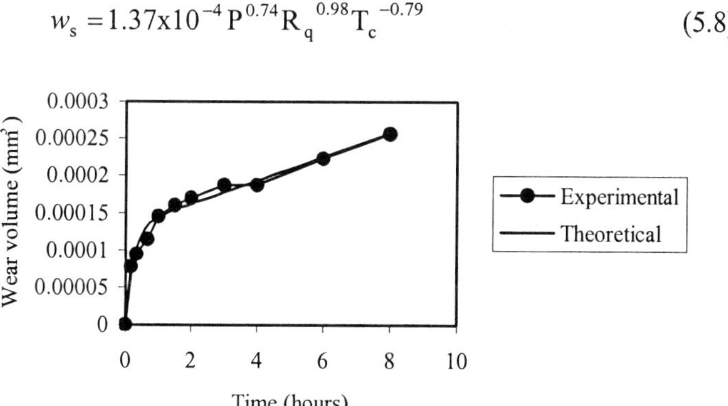

Fig. 5.6. Comparison of experimental wear volume with the theoretical (regressed) curve obtained at 20 N load, 0.35 μm roughness and 150°C.

where

w_s = steady state wear rate, mm^3/hour
P = load, N
R_q = rms roughness, μm
T_c = contact temperature, °C

This equation shows that the steady state wear was clearly influenced by roughness, and temperature as well as load. It is of interest to see that the exponent of temperature is negative. Thus wear rate decreases with temperature in this case. Steady wear rate also decreases as the initial roughness decreases. The strong influence of initial roughness was unexpected as the final scar roughness was nearly the same for all three initial roughness values. It is normally considered that initial roughness mainly affects the running-in part of the wear only. In fact the investigation done with regard to roughness had a practical aim of assessing whether initial roughness influenced steady state wear. This is of importance in engines with regard to the life of the liner. The empirical relation shows the system specific wear behaviour and the need for such modelling. None of the effects can be predicted by the existing theories and the response to wear is specific to the system. While the empirical relationships are of relevance to practice, study of post wear films and wear particles under different operating conditions is necessary to strengthen the theory-practice interface.

Another matter of importance is to decide when the running-in is complete. A possible practical criterion is to consider that running-in is complete when 95% of wear rate at a given point equals the steady state rate. The running-in period was obtained on this basis and was related to the variables considered. Similarly the initial wear rate was also related to the variables involved. These relationships are not given here. It is of interest to note that the running-in time varied from 0.65 to 4.42 hours depending on the operating conditions within a total run of eight hours. In repeat tests it was observed that the repeatability of initial wear rate is poorer in comparison to the steady state wear rate. This may be attributed to the variability involved in the initial wear of the point contact. Thus arbitrary criteria used to determine steady state wear in laboratory machines are unacceptable. It may be argued that it is impractical to conduct long duration tests. But then one has to at least keep in mind that significant errors are possible in short duration tests. Such realisation will help in ameliorating the present procedures as will be discussed in chapter 7. It may be noted that in the present situation the running-in involves basically a change over from point contact to area contact leading to steady state. In real systems running-in refers to micro level adjustments. The present situation is treated as a running-in process in the sense that the evolution of wear depends on the initial contact conditions that include load, temperature and roughness. The methodology developed here is general and is applicable to wear study in any machine and test geometry. The empirical relationship amounts to a wear map. The observed relationships can be represented graphically where the influence of parameters can be effectively visualised. An unpublished example utilising Surfer 7.0 software is given in Fig. 5.7. Wear rate as a function of roughness and temperature at a load of 20 N is shown in (a), while (b) shows the wear behaviour in terms of intensity ranges. The wear rates are given in mm^3/hr. For comparison purposes the overall wear coefficient $K^{/}$ will be useful. The $K^{/}$ values ranged from $7.19x10^{-9}$ to $3.9x10^{-8}$ in these experiments. These relatively low wear coefficients are typical of antiwear action. It may be observed that distinction at such low wear rates is effective through the adapted procedure. Limited analysis of films indicated their thickness ranged from 0.06 to 0.12 μm that is again typical for this additive.

No detailed film analysis was carried out. The empirical relations can also lead to more realistic approaches to wear mechanisms. For example the influence of temperature on steady wear rate may be reconciled with the possibility of more protective glassy films as temperature is increased. The wear of the underlying reaction film will now depend on its composition and the extent of contact through the glassy film. Further progress is possible only by detailed film studies as discussed earlier. The conventional model would predict increasing wear with exponential temperature dependence. The analysis was not done in terms of absolute temperature, but it is clear from the relationships that the reality of wear in the present system is altogether different. This is unlike the case for the empirical dry wear model discussed in section 3.7.2. In this case wear rate depended exponentially on the absolute temperature.

5.5 Failure of boundary lubrication

Boundary lubricant layers will fail when conditions are severe enough. This results in significant oxide/metal contact leading to large scale adhesion. Such a transition leads to what is commonly referred to as scuffing. From a practical point of view scuffing may be defined as a situation resulting in unacceptable surface damage and constitutes a failure of the component. In some cases the scuffing may be severe enough to lead to seizure. It is convenient to start the consideration of scuffing with a mineral base oil and then go on to a consideration of the influence of additives.

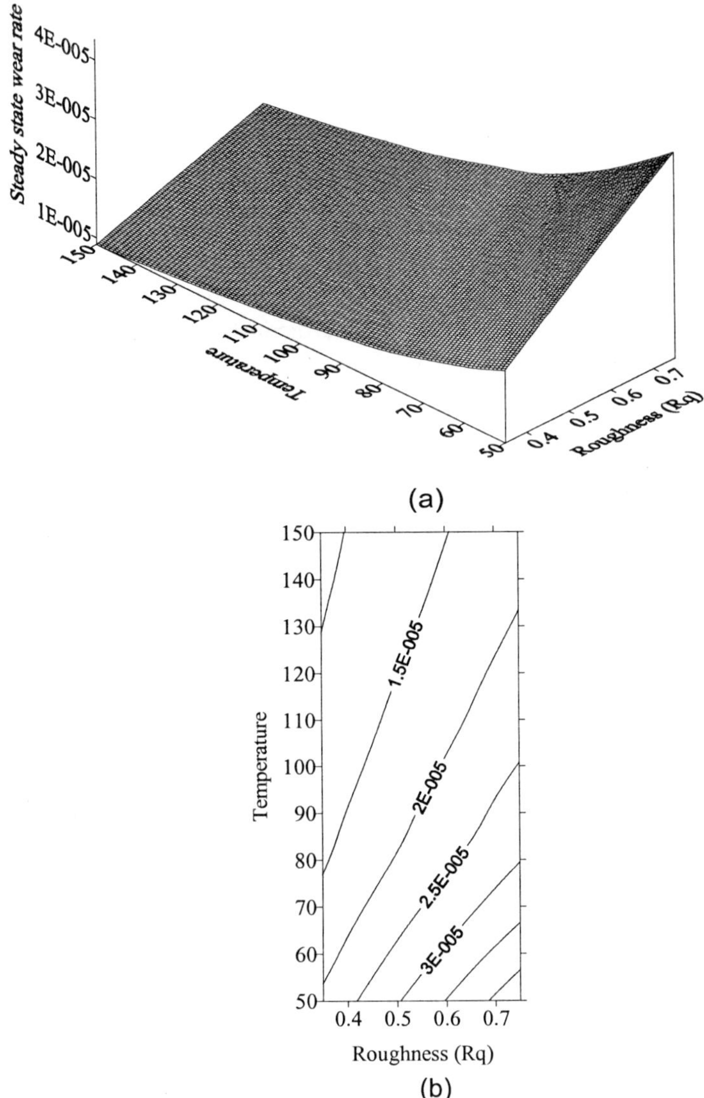

Fig. 5.7. 3-D representation of wear behaviour: (a) Wear rates as a function of roughness and temperature at 20 N load and (b) Wear behaviour shown in intensity ranges.

5.5.1 Scuffing with mineral oils

In boundary lubrication the failure of a mineral lubricant will result in a large scale contact on the thin oxide films at the asperities. The oxide thickness was that obtained under lubricated conditions with relatively low shear stresses and limited contact through films. Also the oxygen availability was limited to the amount dissolved in the lubricant. When lubrication fails there is high friction and correspondingly high shear stress leading to the removal of oxide layer. While oxidation should increase due to higher temperatures, it cannot keep pace with the removal rate leading to metallic contact and scuffing. Hence the hypothesis that lubrication failure leads to scuffing is reasonable with mineral oils. It is common to refer to 'lubrication failure' as 'lubricant failure' and this terminology is adopted from now on. The prediction of lubricant failure is at present difficult. One possibility is that failure occurs when film defect reaches a critical value. It may be recalled from chapter 3 that this criterion was used to model failure temperature of fatty acids and other polar compounds as a function of concentration. However the actual failure temperatures observed were generally low in these tests. For example the failure temperature observed by Frewing [25] for 1% stearic acid in white oil was as low as 60°C. Grew and Cameron [26] observed scuffing temperatures ranging between 50 and 160°C for hexadecylamine in n-hexadecane for different concentrations and loads. On the other hand large number of researchers have reported that mineral oils alone without additives have failure temperatures around 150°C [27,28]. As pure hydrocarbons should fail at lower temperatures it is considered that higher failure temperatures are due to sulphur and oxygen containing impurities in the base oil. There is also clear evidence of the role of polar impurities in fuels. One important problem faced in aircraft engines was the failure of piston pumps observed with hydrotreated aviation turbine fuels briefly considered in chapter 2. The fuel pumps are lubricated by the fuel itself and the impurities act as boundary lubricants. Their removal through hydrotreatment created the tribological problem of high wear and in some cases seizure. It was difficult to understand the relative importance of different impurities which included polycyclic aromatics, oxygenated compounds, and sulphur compounds [29]. The present solutions include change of piston pump materials as well as incorporation of ppm level of additives. The additive approach is limited to military aircraft. Similar problems exist with low sulphur (<0.05%) diesel fuels with the fuel pumps prone to high wear and scuffing problems [30]. In all such cases test methods at laboratory level become necessary. Such procedures, which are empirical, take a long time to develop. Validation is attempted by a comparison between the test parameters in the developed rig and fuel pump. The laboratory test procedures are being developed on a trial and error basis [31].

The criterion for mineral oil failure first proposed by Blok [32] and used with modifications to date is the idea of critical failure temperature. He considered lubricant failure occurs at around 150°C which is the sum of the bulk and flash temperatures. The flash temperature considered was the maximum temperature. For rolling/sliding contacts the equation proposed by Winer [33] in SI units is

$$T_f = \frac{1.11 f\, W|U_1 - U_2|}{(\beta_1 U_1^{1/2} + \beta_2 U_2^{1/2})} \frac{1}{l\omega^{1/2}} \tag{5.9}$$

where

T_f	= flash temperature rise °C
f	= friction coefficient
ω	= instantaneous width of band contact
W	= instantaneous load in conjunction
l	= instantaneous contact length
U_1, U_2	= instantaneous velocities of surfaces 1 and 2 tangential to the conjunction zone and perpendicular to the band length
$\beta_{1,2}$	= thermal contact coefficients of 1 and 2 with $\beta = \sqrt{k \rho c}$ where k, ρ and c are the thermal conductivity, density and specific heat

This equation considers the maximum temperature rise based on Hertzian contact dimensions. The equation follows the same methodology as used to calculate the temperature rise over the geometric area with the difference that the heat is now flowing into the two surfaces which are both moving relative to the source. Moving source theory is applied to both the surfaces assuming Peclet numbers exceed 5.0 for both the surfaces which is normally the case for gears. This failure criterion ignores the effect of viscosity. Instantaneous values are considered to accommodate the general case of line contacts with variable curvatures.

The main interest of Blok was the scuffing of gears that involve rolling/sliding contacts. Such contacts involve EHD lubrication and any failure should involve the failure of the EHD film resulting in the boundary regime and eventual scuffing. Hence viscosity which determines the film thickness must be an additional parameter to be taken into account. Other issues involved are the estimation of bulk temperatures and friction coefficient. A large effort in this area has finally led to gear scuffing criteria that are easy to use. These are discussed by Enrrichello [35] and briefly considered here. The contact temperature is expressed as

$$T_c = T_b + T_f$$

where T_b and T_f refer to the bulk and flash temperatures

The flash temperature rise is obtained on the basis of Eq. (5.9) taking into account the load sharing between gears. The friction coefficient is obtained empirically on the basis of surface roughness. The bulk temperature that is applicable is again obtained empirically on the basis of inlet temperature at a selected speed.

The recommended failure temperatures in °F are as follows

$T_s = 146 + 59\, ln(V_{40})$ for mineral oils without anti-scuff additives and (5.10a)

$T_s = 245 + 59\, ln(V_{40})$ for the oils containing anti-scuff additives (5.10b)

At a fundamental level the issue to be resolved is the failure criterion for thin films taking into account the roughness effects. For this purpose effective modelling of mixed lubrication is necessary. The complexity involved in mixed lubrication shall be considered in chapter 8. Some models are available which consider that scuffing is essentially related to EHD film failure [35,36]. The concept is that when film thickness is reduced to a critical value the asperity contact through such films leads to failure. These models are not considered here in detail. As boundary lubrication effects are not known such models based on EHD film thickness alone are not complete. It may be argued that film thinning is a necessary but not sufficient criterion for scuffing. The model has been successful when mineral oils of different viscosities were tested in FZG gear rig [36]. The success can be due to the fact that the critical temperature for thinning are higher than the adsorption related failure temperatures of mineral oils. As stated earlier the mineral oils have typical failure temperature of about 150°C under boundary lubrication conditions. The failure temperatures ranged from 120°C to as high as 420°C depending on the oil viscosity in the gear tests. With some exceptions these temperatures are higher than boundary failure temperature. It is difficult to separate the relative influence of film thinning and asperity contact. Understanding shear failure of both boundary and thin EHD films without asperity contact can form a useful basis for such a study. This can be done only with surfaces of nano level roughness. Such studies on film failure are not available.

Scuffing is considered rather simply as that process which results in unacceptable surface damage. Scuffing has been defined by the Institution of Mechanical Engineers [37] as "gross damage characterised by the formation of local welds between surfaces". OECD [38] defined it as "localised damage caused by the occurrence of solid-phase welding between scuffing surfaces, without local melting". Ludema [39] considered scuffing to be due to roughening of surfaces by plastic flow whether or not there is material loss or transfer. The possibility of local welding at asperity level is always there due to adhesion and is invoked in modelling adhesive wear. Gross damage leading to scuffing can only occur when adhesive growth in contact area propagates to a level that amounts to gross transfer of material. In some cases there can be an *increase* in the weight of one of the surfaces due to material transfer when scuffing occurs. The exact mechanism by which a transition occurs from local (asperity) level to gross damage is not well understood and continues to be an area of investigation. In dry scuffing tests a recent carefully conducted work [40] suggests that scuffing of aluminium alloys is due to sub-surface plastic failure. Yet another work conducted in vacuum in dry conditions [41] and low temperatures considered that there is a critical growth of the real area beyond which scuffing occurs. Relevance of these studies to lubricated contacts is difficult to predict at this stage. Different definitions arise as there is no clearly accepted model for scuffing. At present the practical way out is to accept qualitatively that scuffing leading to gross damage does occur beyond a level of severity. Experimentally based semi-empirical models offer the best guidelines available to predict scuffing.

5.5.2 Control of scuffing – EP additives

The role of EP additives and the generally accepted mechanisms have been already covered in chapter 4. When mineral oils fail to lubricate and there is a tendency to scuff the EP additives take over. At elevated temperature they react swiftly forming a reaction film. This reaction film prevents metal-metal contact and hence avoids scuffing. This is schematically illustrated in Fig. 5.8 in which the arrows indicate scuffing load. This can be observed in any machine that can be loaded to the level of EP film failure. Increasing temperature here is a consequence of increasing severity of operating conditions. The influence of polar compounds which can provide some additional protection in comparison to mineral oil is also illustrated in the diagram. It may be seen that the EP additive becomes active at a threshold temperature when it reacts effectively on the surface. This means the additive will interact only when necessary and thus avoid unnecessary reaction and wear. It is of interest to consider what happens beyond initial film formation. As the temperature increases due to severity the reaction rate should increase causing increased film wear as discussed in the section on chemical wear. The assumption here is that unlike the case with antiwear additives the strong reaction leads to the conventional film wear where relatively thick films wear out and the overall reaction follows the parabolic law. It is also possible that films may change and offer an increased

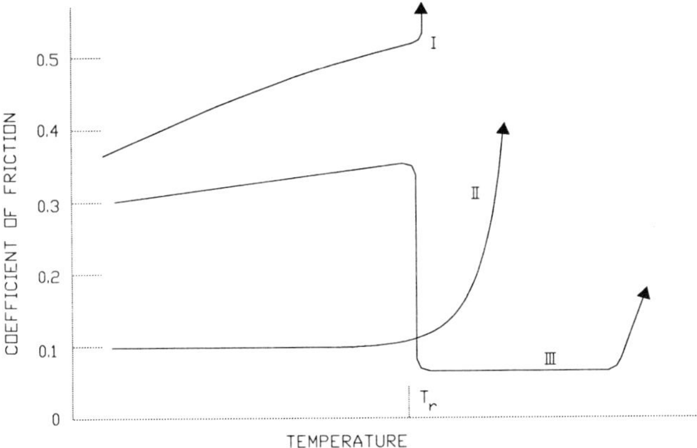

Fig. 5.8. EP action of additive (III) with threshold reaction temperature T_r as compared with base oil (I) and fatty oil additive (II). Arrow represents scuffing.

barrier to reaction reducing the wear rate. Finally the severity can reach such a level that EP films fail. The failure is normally attributed to an imbalance between the removal rate and the reaction rate. Neither of them can be modelled with any certainty. As increased temperature should increase the reaction rate exponentially it is difficult to prove such a hypothesis on the basis of reaction rate in one direction alone. Also there should be an adequate reason to explain the change over to a high removal rate. The possibility is that at high temperatures, significant film modification mechanisms set in, altering the film composition at the surface and sub-surface level. This can cause a weakening of the film leading to a removal rate that is higher than the rate of formation. Formal approaches based on solid state chemistry can clarify these mechanisms. If the film does not fail by this mechanism it may eventually fail by melting. Another possibility is the softening of the underlying metal due to high temperatures leading to film collapse [14]. The possibility of film melting is difficult to envisage for sulphide-oxide films as iron sulphide has a melting temperature in excess of 1000°C. The additive concentration is another factor of importance. It is known that load carrying capacity increases with concentration up to a point beyond which the influence becomes negligible. The net reaction rate reduces with lower concentrations and can lead to lower failure load. Another complicating factor is the nature and extent of oxidation that changes the film composition. The iron oxides that form are highly complex and can be a combination of several oxides. While it is known that oxides are a necessary component for load carrying capacity their role has not been studied in detail. A recent study shows that dissolved oxygen concentration can influence significantly the load bearing capacity of dibenzyl and diphenyl disulphides [42]. The mechanisms are complex and it is difficult to establish any guidelines in this area.

One important qualitative aspect has to be considered in relation to EP action. This is related to the rate at which severity is increased. When the rate is low the films will have a particular composition. When the rate is high the films are ill conditioned and will have different composition. Scuffing condition for the two cases will be different. This translates into different severity levels for failure and is not brought out in the schematic diagram. These considerations are of importance in EP evaluation and will be discussed in chapter 7.

The above discussion shows that EP action of an additive cannot be modelled effectively. Progress in understanding is possible only through a detailed study of worn films up to a stage prior to scuffing in the laboratory machines. Reaction rate studies on the post wear films can provide realistic estimates of kinetic parameters. These studies can also clarify how the EP films evolve in progressive running and their role in failure.

Antiwear additives can also provide an increased load carrying capacity as compared to base oils alone. These additives are slow acting in comparison to EP additives. Some of these additives are called mild EP additives. Wherever necessary a combination of EP and antiwear additives is used. Their mechanism can only be discussed in general terms as one of combined EP and antiwear action.

Scuffing control is also possible through material modifications. One example is sulphide-nitride coatings obtained by simultaneous diffusion of nitrogen and sulphur. Another example is piston rings for which a range of coatings are available that include chrome plating, molybdenum coating, and gas nitriding. One important role of piston ring coatings is to control scuffing during running-in in addition to wear control. Detailed consideration of tribological coatings is available in several sources. A recent review [43] may be consulted for information.

5.5.3 Practical aspects

The practical formulations in industry for EP and antiwear action can be combinations of one or more additives. The basic technology involved is to have the required EP activity while at the same time controlling the wear. Industrial systems operate under varied conditions. EP action will be needed only when there is an excursion to severe operating conditions. For the rest of the time it is the low wear that is important. Any excessive action of the EP additive can lead to large wear that is unacceptable. This balancing is achieved by controlling the activity levels of the additives involved. From a consideration of the antiwear mechanism proposed earlier, effective films offer a strong barrier to chemical reaction. Such films can interfere with quick reaction of EP additives when needed. The author has not come across detailed mechanistic studies of this aspect in the literature. One example of a complex formulation is the lubricant used for hypoid rear axles that involve a high degree of sliding. Strong EP additives control the high speed and shock load scuffing tendency. Another need is to control the ridging and rippling under low speed and high torque condition due to plastic deformation. Phosphorous additives are used to control this problem. It is also to be noted that EP additives cannot be used in some systems. For example EP additives cannot be used in engine oils due to their strong corrosive action on copper based bearing materials and other components. Only the slow acting dithiophosphate has been found to be a suitable additive. The scuffing control during running-in is hence more difficult and the variety of surface coatings used assist in the control of scuffing. The possible control due to the slow acting dithiophosphate cannot be relied upon.

The modelling possibility of EP action is limited. The present approach is to evaluate scuffing in laboratory machines and this is considered in chapter 7. Such empirical testing has its limitations and EP action can only be ensured through final evaluation in the real system. It is also to be realised that the additives generate reaction products during use and their influence on the overall system also needs careful consideration. Environmental issues related to disposal of such products is also becoming more important.

References

1. C. N. Rowe, Some aspects of the heat of adsorption in the function of a boundary lubricant, ASLE Trans., 9 (1966) 101.
2. T. A. Stolarski, A system for wear prediction in lubricated sliding contacts, Lub. Sci., 8 (1996) 315.
3. T. A. Stolarski, Adhesive wear of lubricated contacts, Trib. Int., 12 (1979) 169.
4. R. G. Bayer, Prediction of wear in a sliding system, Wear 11 (1968) 319.
5. A. Beerbower, A critical survey of mathematical models for boundary lubrication, ASLE Trans., 14 (1971) 90.

6. C. N. Rowe, Role of additive adsorption in the mitigation of wear, ASLE Trans., 13 (1970) 179.
7. C. N. Rowe, Lubricated wear, in M. B. Peterson and W. O. Winer (eds.), Wear Control Handbook, ASME, New York, 1980, 143-160.
8. K. Nakayama and T. Sakurai, The effect of surface temperature on chemical wear, Wear, 29 (1974) 373.
9. J. Molgaard, A discussion of oxidation, oxide thickness, and oxide transfer in wear, Wear, 40 (1976) 277.
10. S. A. Bradford, Fundamentals of corrosion in gases, ASM Handbook, Vol. 13, 1987, 61-76.
11. A. B. Vipper, A. K. Karaulov and O. A. Mischuk, New data on the mechanism of antiwear action of zinc dithiophosphates in lubricating oils, Lub. Sci., 7 (1994) 93.
12. E. P. Becker and K. C. Ludema, A qualitative model of cylinder bore wear, Wear, 225-229 (1999) 387.
13. G. C. Barber and K. C. Ludema, The break-in stage of cylinder-liner wear: A correlation between fired engines and laboratory simulator, Wear, 118 (1987) 57.
14. A. Sethuramiah, H. Okabe and T. Sakurai, Critical temperatures in EP lubrication, Wear, 26 (1973) 187.
15. M. Lukas and D. P. Anderson, Laboratory used oil analysis methods, Lub. Eng., (1998) 31.
16. W. W. Seifert and V. C. Westcott, A method for the study of wear particles in lubricating oil, Wear, 21 (1972) 22.
17. M. Hallouis, M. Belin and J. M. Martin, The role of sulphur in ZDDP induced reaction films formed in the presence of ZDDP: Contribution of electron spectroscopic imaging technique, Lub. Sci., 2 (1990) 337.
18. J. M. Martin, C. Grossiord, Th. Le. Mogne, S. Bec and A. Tonck, The two-layer structure of Zndtp tribofilms: Part I: AES, XPS and XANES analyses, Trib. Int., 34 (2001) 523.
19. A. V. Sreenath and N. Raman, Running-in wear of compression ignition engine: Factors influencing the conformance between cylinder liner and piston ring, Wear, 38 (1976) 271.
20. P. Pawlus, A study of the fuctional properties of honed cylinder surfaces during running-in, Wear, 176 (1994) 247.
21. K. C. Joseph and N. Raman, Fractal characterisation of running-in behaviour of an IC engine piston ring and cylinder liner combination, in Harprasad (ed.), Proc. Second Int. Conf. On Industrial Tribology, Dec 1999, Hyderabad, India, 247.
22. M. Zheng, A. H. Naeim, B. Walter and G. John, Break-in liner wear and piston assembly friction in a spark ignition engine, Trib. Trans., STLE, 41 (1998) 497.
23. Y. Z. Hu and K. Tonder, Application of a dynamic system model for running-in, Proc. Int. Conf. On Wear of Materials, ASME, 1991, 201.

24. R. Kumar, B. Prakash and A. Sethuramiah, A systematic methodology to characterise the running-in and steady state wear process, Wear, 252 (2002) 445.
25. I. J. Frewing, The heat of adsorption of long-chain compounds and their effect on boundary lubrication, Proc. Roy. Soc. London, Series A, 182 (1944) 270.
26. W. J. S. Crew and A. Cameron, Thermodynamics of boundary lubrication and scuffing, Proc. Roy. Soc. London, Series A, 327 (1972) 47.
27. R. M. Matveevsky, The critical temperature of oils with point and line contact machines, Journal of Basic Engineering, ASME, 89 (1965) 754.
28. E. F. Leach and B. W. Kelly, Temperature, the key to lubricant capacity, ASLE Trans., 8 (1965) 271.
29. J. Appeldoorn and W. G. Dukek, Lubricity of jet fuels, SAE paper 660712 (1966) 428.
30. D. Wei and H. A. Spikes, The lubricity of jet fuels, Wear, 111 (1986) 217.
31. D. Cooper, Laboratory screening tests for low sulphur diesel fuel lubricity, Lub. Sci., 7 (1995) 133.
32. H. Blok, Theoretical study of temperature rise at surfaces of actual contact under boundary lubrication conditions, Proc. Inst. Mech. Engrs., London, 2(1937) 471.
33. W. O. Winer and H. S. Cheng, Film thickness, contact stresses and surface temperatures, in M. B. Peterson and W. O. Winer (eds.), Wear Control Handbook, ASME, 1980, 121-139.
34. R. Enrrichello, Friction, lubrication and wear of gears, in P. Blau (ed.), Friction Lubrication and Wear Technology, ASM International, 1992, 535-545.
35. S. C. Lee and H. S. Cheng, Correlation of scuffing experiments with EHL analysis of rough surfaces, J. Trib., ASME, 113 (1991) 318.
36. J. Castro and J. Seabra, Scuffing and lubricant breakdown in FZG gears Part I. Analytical and experimental approach, Wear, 215 (1998) 104.
37. Memorandum on Definitions and, Symbols and Units, Proc. I. Mech. E., 4, 1957.
38. Glossary of Terms and Definitions in the Field of Friction, Wear and Lubrication, OECD, Research Group on Wear of Materials, Paris, 1969, 3.
39. K. C. Ludema, A review of scuffing and running-in of lubricated surfaces, with asperities and oxides in perspective, Wear, 100 (1984) 315.
40. T. Sheiretov, H. Yoon and C. Cusano, Scuffing under dry sliding conditions-Part II: Theoretical studies, Trib. Trans., STLE, 41 (1998) 447.
41. Q. Quyang and K. Okada, A study on the quantitative description of the seizure behaviour of steels at low twmperature in a vacuum, Trib. Trans., STLE, 41 (1998) 301.
42. T. Murukami and H. Sakamoto, Effect of dissolved oxygen on lubricating performance of oils containing organic sulfides, Trib. Int., 32 (1999) 359.
43. K. Holmberg, A. Matthews, and H. Ronkainen, Coating tribology-contact mechanics and surface design, Trib. Int., 31 (1998) 107.

Nomenclature

a	an integration constant
a_1	constant
b	non-linear coefficient having inverse relation with running-in period
β_1, β_2	thermal contact coefficients of 1 and 2 with $\beta = \sqrt{k \rho c}$ where k, ρ and c are the thermal conductivity, density and specific heat
A_e	elastic part of real area
A_p	plastic part of real area
A_r	total real area
C'	constant
C	additive concentration, mole fraction
E	heat of adsorption
f	coefficient of friction
H	hardness
k_e	rate constant in logarthimc growth law
k_m	wear coefficient for metal contact as defined in the text
k_n	rate constant in parabolic growth law adapted to number of cycles
K'	overall wear coefficient in chemical wear
K_b	boundary wear coefficient in adsorption based models
K_f	film wear coefficient (inverse of number of cycles needed to form critical thickness)
l	sliding distance
l	instantaneous length in line contact
n	number of cycles
P	load
Q_p	activation energy
R	molar gas constant
R	coefficient of determination
R_q	rms roughness
t	time
t_0	fundamental time of vibration of a molecule
t_0'	ratio of fundamental time of vibration of additive and base fluid molecules
u	sliding velocity
T_b	bulk temperature

T_c	critical contact temperature
T_f	flash temperature rise
T_r	threshold reaction temperature
T_s	temperature of the surface
V	wear volume
$\dot{V}$	wear rate expressed as wear volume per unit time at time t
$\dot{V}_o$, $\dot{V}_s$	wear rate at time zero and steady state
U_1, U_2	instantaneous velocities of surfaces 1 and 2 tangential to the conjunction
W	load
W	instantaneous load in conjunction
w_s	steady state wear rate expressed as wear volume per unit time
x	diameter of area associated with an adsorbed molecule

Greek Letters

α	fractional contact area as applicable
ω	instantaneous width of band contact
γ	$=(1+3f^2)^{0.5}$
ΔE	*difference* in heat of adsorption of additive and base fluid
Δr	film thickness
ΔS^0	overall entropy change

6. Wear of non-metallic materials

6.1 Introduction

The use of non-metallic materials in tribological applications is of growing importance. The major classes of materials used are polymers and ceramics and the main types used have been covered in chapter 1. Another class of materials being developed is metal matrix composites in which the metal matrix is modified by incorporating fibres and /or particles. The investigations being done in these areas are very extensive and several specialized books are available, some of which were cited in chapter 1. Both dry and lubricated wear of polymers and ceramics are covered in this chapter. Coverage of the material in one chapter is necessarily limited. The scope of the chapter is confined to a broad appreciation of the mechanisms involved and the issues related to applications. Relatively more emphasis is laid on lubricated wear.

The first section considers the main wear modes of commonly used polymeric materials and composites in dry wear. The second section deals with the lubricated wear of polymers. This section includes the application aspects and observations on the lubricated wear. The observations are meant to comment on the missing links using the available information. The dry and the lubricated wear of ceramics form the third and fourth sections. The lubricated wear of ceramics again includes sub-sections dealing with the application and observations on lubricated wear. Sequencing of the sub-sections is different for polymers and ceramics. Abrasive wear is excluded from the coverage.

6.2 Dry wear of polymers

Polymer tribology has been covered at an introductory level in section 1.5.2. To recall the main polymers used are high density polyethylene (HDPE), polyamide (Nylon 6-6), polyoxymethylene (Acetal), and polytetrafluoroethylene (PTFE). The repetitive units from which these polymers are made are shown in Fig. 6.1. In this figure the repetitive unit is the same for normal low density polyethylene (LDPE) and HDPE. The major difference between these two polymers is with regard to the molecular chain structure. In HDPE the molecular chains are linearly ordered while in the case of LDPE the chains have a degree of random orientation. The polyetheretherketone (PEEK) and polyimide are also included in this figure. These two polymers are used for high temperature applications. In particular PEEK is gaining importance for such applications. Wear of pure polymers is considered first followed by polymer composites.

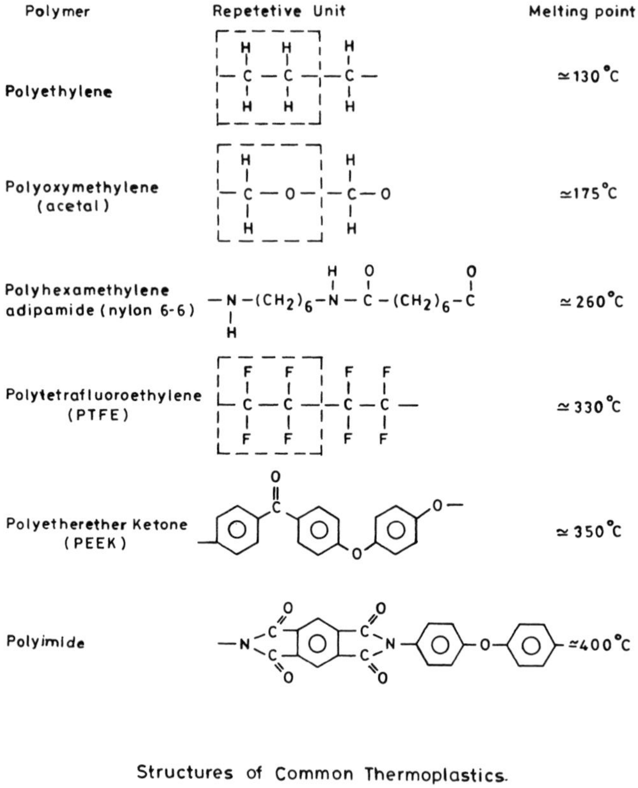

Polymer	Melting point
Polyethylene	≃130 °C
Polyoxymethylene (acetal)	≃175 °C
Polyhexamethylene adipamide (nylon 6-6)	≃260 °C
Polytetrafluoroethylene (PTFE)	≃330 °C
Polyetherether Ketone (PEEK)	≃350 °C
Polyimide	≃400 °C

Structures of Common Thermoplastics.

Fig. 6.1. Polymer structures.

6.2.1 Wear of pure polymers

The main mechanism postulated for polymer wear may be called the transfer mechanism. This mechanism refers to the situation where the counter face is a hard surface in comparison to the polymer. Usually polymer applications involve a hard metallic counter face and so the majority of studies were done with smooth metal counter faces and in some cases against glass surfaces. Polymer-polymer tribology has also been studied, though to a lesser extent. The transfer mechanism involves the following sequence:

1. As sliding starts polymer material is transferred to the counter face.
2. With further sliding the transferred material gets oriented in the direction of sliding. Surface roughness helps in anchoring the transferred layer.
3. The transferred material is removed after further sliding as loose debris. The debris particles can also agglomerate resulting in large clustered particles.
4. There can be back transfer to the pin that can influence the above processes.

The main difference between the various polymers is in the nature of transfer. Polymers with smooth molecular profiles tend to have low friction and thin transfer films as discussed in section 1.5.2. Thus both HDPE and PTFE transfer thin films onto the counter face. The wear of pure PTFE is high because the films are weakly attached to the counter face. In comparison the wear of HDPE is much lower. Nylon and other branched chain polymers involve lumpy transfer and relatively high friction. Their wear resistance depends on the strength of adhesion to the counter face. For example the wear of LDPE is much higher than nylon due to its poor adherence. The above observations are directional and generally accepted. At a fundamental level many questions remain. The detailed crystalline structures of the polymers are firstly very different. For example PTFE has a banded structure with alternating crystalline and amorphous regions. HDPE has a spherulite structure. Such aspects have to be taken into account for a fundamental understanding of transfer. The removal process also needs to be explained effectively from basic considerations. Fundamental studies in these areas are continuing and the interested readers can consult the available literature [1,2,3,4].

There is evidence that HDPE can undergo fatigue wear [5] beyond a certain number of stress cycles. Thus other wear modes besides transfer are also possible. Some work reported on PTFE wear in a three pin-on-disk machine by Agarwal et al [6] showed structural changes. These tests were conducted at speeds of 0.73 and 1.47 m/s at two different loads of 44 and 74 N. The wear particles showed higher crystallinity as compared to the bulk polymer based on X-ray diffraction. On the other hand the overall pin surface showed lower crystallinity as assessed by multiple internal reflection (MIR) spectroscopy. The original polymer had a crystallinity of 85% and the increase in crystallinity of wear particles was up to

5%. The reduction in crystallinity at the surface was up to 8%. On the pin surface in some localised zones the removal process was probably due to fatigue. Increases in crystallinity of the order of 30% in wear particles were found for ultra high molecular weight polyethylene (UHMPE) by Marcus et al [7] in sliding tests with stainless steel using distilled water as the lubricant. Several other examples of structural changes at the surface and subsurface are available with different polymers. These variations complicate the understanding of wear mechanisms.

Wear rate for polymers is defined in terms of the specific wear rate volume/(sliding distance x Load) and has the SI units of m^3/Nm. Attention may be paid to the units as sometimes the wear volume is expressed in mm^3. This approach is adopted because hardness is ill defined for viscoelastic materials. A dimensionless wear coefficient, commonly used for metals, is rarely used. Data on wear rate for polymers is recorded by several authors with different machines and operating conditions. Such data can be used for rough screening of candidate materials. Some examples will be considered in the next section.

Quantitative wear models are limited for polymers. Jain and Bahadur [8] modelled wear on the basis of fatigue. Fatigue was related to the cyclic tensile stresses acting at the asperities. Fatigue is one of the mechanisms, and such modelling can be effective only when fatigue wear is dominant. There is evidence for subsurface fracture and removal of thick flakes observed by some researchers. Modelling in such cases has been attempted by fracture mechanics approaches [9]. A more recent effort to model wear is based on dimensional analysis [10].

The roughness influence of the counter face has been studied by several researchers. The wear is minimum at an optimum roughness below and above which the wear tends to increase. Such behaviour is also observed with regard to friction. Czichos [11] and several others proposed that the higher wear and friction with increasing roughness is due to the increased abrasive wear of the polymer. The increasing friction and wear below an optimum roughness is more difficult to explain. The reason advanced is that with very smooth surfaces the transferred layers are more easily removed. This is because of the lack of mechanical anchoring provided by the roughness. The optimum roughness is in the range of 0.2 to 0.4 μm. When the roughness is optimum, the roughness is just covered by the transferred polymer with minimum possibility for abrasion. With metal-metal contacts also when roughness is very low, a sudden increase in friction and damage can occur. Unlike the case with polymers this effect is due to strong adhesion. The effect was experimentally demonstrated by Hirst and Hollander [12] for boundary lubricated steel surfaces. Strong adhesion occurred when the surface roughness σ

was very low with typical values of less than 0.02 μm. This effect is probably due to the increased interatomic forces between the very smooth surfaces.

In polymer friction, unlike in metallic contacts, the roughness of the hard counter face is unlikely to change. High peaks that can be approximately characterised by 10- point height (or other measurements) may persist and continue to abrade the polymer. 10-point height refers to the distance between the average height of the five highest asperities and the five lowest valleys in a selected sampling length. The sharpness of asperities, as determined by slopes is also of importance since sharper asperities will cause more abrasion. In the case of polymer composites the influence of fillers on roughness also has to be taken into account.

Polymer-polymer friction has been investigated by Czichos [11]. He found that work of adhesion was reasonably correlated to friction for several polymer pairs. As work of adhesion can be found more accurately for polymers this is direct evidence of the role of adhesion on friction. With metallic surfaces it is difficult to define adhesion as discussed in chapter 1 and such correlations are not possible.

Most of the research in dry polymer wear was conducted with sliding contacts. More attention is now being paid to other situations that involve rolling/sliding and impact [13,14]. Significant attention is also being paid to abrasive wear of polymers. One important material that is emerging for low wear applications in this area is PEEK.

6.2.2 Wear of polymer composites

Polymer composites may be defined as those materials in which particles and/or fibres of different types are incorporated in the polymer matrix. Such fillers serve two purposes. Firstly the fillers can increase the strength properties. Many fillers also improve the tribological properties. In some cases polymer-polymer composites can also be used. One example is the blending of PTFE with PEEK to reduce friction. The composites may be classified into particulate and fibre composites. Fibre composites consist of reinforcing fibres. These composites are receiving increased attention. The normally used fibres include carbon, glass, and aramide. The composites can have randomly oriented short fibres or directionally oriented long fibres. The performance of such composites depends on their orientation with respect to the sliding direction. The stress distributions within the fibres and at the fibre-polymer matrix are very important in the wear behaviour of these composites. The developments are available in the literature and some selected references may be cited [15,16,17,18].

The most extensively used composites are PTFE composites and emphasis will be placed on these materials. A wide variety of fillers are used that include bronze, graphite, MoS_2, lead sulphide, copper sulphide, and copper oxides. The particles can have a wide range of sizes and morphologies. In many cases more than one filler is used with the desire to control friction as well as wear. A wide variety of materials are commercially available. The fundamental question is what filler should give the best performance. Unfortunately the action mechanisms of fillers are not well understood. The possible mechanisms for wear reduction postulated may be classified as follows:

1. Development of filler enriched layers on the polymer as well as on the counter face. The fillers in the enriched layers support part of the load.
2. The polymer transfer mechanism is inhibited by the particles at the surface and in the bulk leading to low wear of the polymer matrix.
3. Inhibition of subsurface crack growth that prevents thick material transfer.

Blanchet and Kennedy [19] have provided good evidence for the inhibition of subsurface crack growth for filled PTFE at higher sliding speeds. While there is no doubt that these mechanisms are important their relative importance depends on the operating conditions. Another important issue is why some fillers are more effective than the others. One interesting observation based on surface analysis is that the fillers were found to be effective when they react chemically with the counter face [20,21]. It is argued that such a reaction will result in strong adhesion and provide a wear resistant surface film. In the overall wear process several other aspects have to be taken into account. It may be considered that chemical reaction is one of the requirements for wear reduction. The number of fillers and their combinations are endless and the idea of chemical reactivity can help in choosing the fillers.

Zhao and Bahadur [22] recently attempted to generalise the concept of chemical reactivity on the basis of free energy change involved in a reaction. It is known that a chemical reaction is feasible when the free energy change is negative. The possibility of reaction increases as the negative free energy increases. The authors have evaluated the available experimental work, and showed that when chemical reactions occurred the free energy change was negative. When reaction did not occur and the filler was ineffective, the possible chemical reaction between filler and metal had a positive free energy change. This idea strictly applies to equilibrium reactions. Nevertheless this is a reasonable approach and provides a criterion for choosing fillers. It is worthwhile to pursue this idea with a detailed analysis of the reaction kinetics. Such solid state reactions are also of importance in EP lubrication as considered in the previous chapter.

The friction and wear behaviour of composite materials as well as pure polymers varies in a complex manner with operating conditions. The viscoelastic behaviour is a function of strain rate and temperature. Theoretical prediction of friction is available for rubber at low sliding speeds [23] and is not considered here. It is difficult to make theoretical predictions on this basis for polymers particularly under severe operating conditions. Reliance has to be placed on the experimental observations. The wear behaviour can best be appreciated from the available literature. One typical example of such a study by Holmberg and Wickstrom [24] is considered here. They have reported the friction and wear behaviour of 22 different materials that are commercially available. Only wear tests are considered here and the wear behaviour is reproduced in Fig. 6.2. The wear was measured in terms of the depth of pin wear in 1000 h. The wear tests were conducted with a multi-pin-on-cylinder tester that measured wear of 12 different pins simultaneously at a sliding speed of 0.1 m/s. The polymers were grouped into five classes that included both pure and filled materials. The polymer groups tested were polyamides, PTFE, polyethylenes, polyesters, and polyacetals. The polyethylene used was UHMWPE with a molecular weight of more than 10^6. The composites used were commercial materials. The tests were conducted at ambient temperature as well as at a low temperature of –35°C. From this figure the large influence of fillers can be clearly seen. Filled PTFE and polyethylene stand out as low wear materials. This fact coupled with their low friction makes them an important group of materials for tribological applications. The wear of pure PTFE can easily be a thousand times higher than the filled materials and so rarely used in the pure form. It is also seen that the low temperature wear behaviour can be very different from the performance at ambient temperature. The difference between the individual polymers in each group is also evident. For example polyamide 66 (nylon 66) showed higher wear than nylon 6. The filler materials studied here are those that are commonly used. Some oil-impregnated polymers were also tested and showed low wear.

The purpose of this example is to show the typical wear behaviour of a wide range of polymers and their composites at a given set of operating conditions. Such studies are directed towards selection of polymers and composites based on their friction and wear behaviour. While available knowledge on mechanisms is helpful, it is not adequate to select materials for specific applications. Several such studies are reported in the literature and the author has selected one example where a large number of materials were evaluated.

6.3 Lubricated wear of polymers

The interest in this section is confined to the influence of conventional liquid lubricants on wear. Solid lubricants like graphite are sometimes used in composites and their influence on wear is not separately considered here. Polymer wear with conventional lubricants has not received much attention. One area that has been extensively studied is the lubrication of UHMWPE used in artificial human joints with water, saline, and serum. Two excellent reviews by Unsworth [25] and Dowson [26] cover the tribological and clinical aspects of these joints. The present chapter does not cover this area but will consider some aspects of this lubrication in chapter 8. The first part of this section considers examples of studies on lubricated wear. Based on this information some observations only could be made regarding mechanisms. It may be observed that the mechanistic studies with liquid lubricants are limited. The next part of the section deals with practical aspects of lubricated bearings.

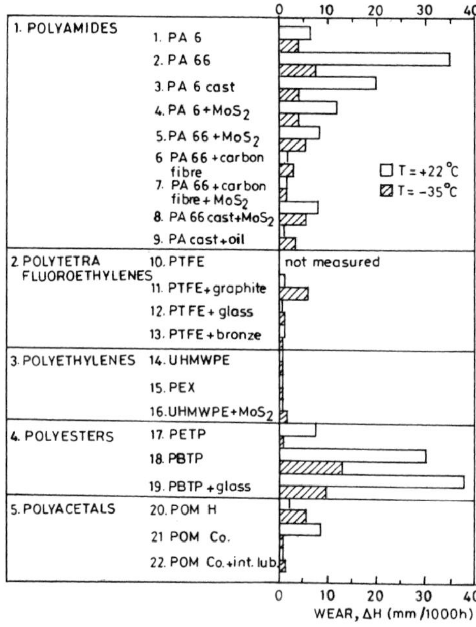

Fig. 6.2. Wear for polymer materials at normal room temperatures and low ambient temperature (p = 5 MPa, υ = 0.1 ms^{-1}). (Reproduced from Ref. [24]).

6.3.1 Experimental observations of lubricated wear

Bramham et al [27] have reported on the influence of contaminants on aircraft plastic bearing liners. The PTFE linings consisting of aramid fabric with different kinds of resins were the materials tested. The tests were conducted in line and area contact. Some liner materials also incorporated glass fibres. The counter face material was stainless steel with a roughness of 0.05 μm. In line contact they found water was particularly deleterious at the high stress level of 60 MPa with regard to wear, while the wear with a hydraulic mineral oil was slightly lower than the dry case. They also found that at lower stress levels the wear rates for both fluids are lower than the dry case. The friction coefficients were always lower than the dry case. There was a time effect also with water and the large increase was observed after several hours of operation in line contact. The comparative wear rates with area contact were lower with the lubricants.

Dickens et al [28] have studied the speed effects on polymer wear with polyphenyl oxide (PPO), PTFE, and PEEK. The authors used a series of polydimethyl siloxane fluids of different viscosities as lubricants. The important observation was that in all cases partial hydrodynamic effects occurred and increased with speed. They persisted even at the low sliding speed of 10^{-3} m/s. The hydrodynamic effects were higher in the line contact. The friction coefficients ranged from 0.01 to 0.1 for PPO and PEEK while they ranged from 0.01 to 0.05 for PTFE. In general friction and wear decreased with sliding speed for all materials. The variations in friction in area contact were lower and though wear clearly decreased with sliding speed they could not be related to the friction variations in a systematic manner. For area contact the friction coefficients ranged from 0.07 to 0.1 for PPO and PEEK while for PTFE they ranged from 0.02 to 0.04.The reduction in the lubricated wear rate was 3-4 orders of magnitude compared to the dry wear at the highest sliding speed of 1.0 m/s. The importance of this work is the clear evidence of hydrodynamic effects. Such effects arise from macro as well as micro deformations. Hence increased hydrodynamic effects are to be expected in line contacts as observed experimentally. The issues related to hydrodynamic effects on wear will be discussed in chapter 8. It may be noted that the term 'hydrodynamic effect' is used in a general sense and covers both hydrodynamic as well as elastohydrodynamic effects.

Another example of lubricated wear studies by Sethuramiah et al [29] may be cited. In this work the friction and wear of pure and graphite filled PTFE were studied in a three pin-on-disk machine. Graphited PTFE was a commercial product containing 15% graphite. The polymer pins were slid against a mild steel disk. The roughness, R_a was 0.2 μm in the radial direction. The tests were conducted in three steps of 30-minute duration and the friction and wear in the final step formed the basis for comparison. The speed range was 0.73-1.47 m/s while the load range was 44 to 164 N. The corresponding contact pressures ranged from 0.44 to 1.64 MPa. The lubricant used was base oil with a viscosity of 46.1 cSt at 40°C. One ml of the oil was spread on the disk surface at the start of each step of 30-minute duration. Between the steps the disk was cleaned before applying fresh lubricant. The wear rates observed in these tests are summarized in Fig 6.3. The range of friction coefficients observed for lubricated PTFE are shown in brackets.

Firstly focusing on the behaviour of pure PTFE it may be seen that wear rate in lubricated contact is two to three orders of magnitude lower than in dry contact. The specific wear rate of dry PTFE ranged from 3.8 to 6.0×10^{-13} m^3/Nm. Speed has a strong influence on lubricated wear. The wear rate was lower at the lower speed of 1.05 m/s as compared to the higher speed of 2.09 m/s. The range of f values for low speed condition was 0.04 to 0.06. For the high speed condition the range was 0.07 to 0.12. At higher speed the increased temperature will reduce the film thickness and increase boundary contact and wear. Thus in the case of PTFE it may be argued that wear is inversely related to the hydrodynamic effects.

Graphited PTFE (GP) showed a complex behaviour. The overall wear rates for this material in the dry case were clearly lower than that of pure PTFE, which can be attributed to the influence of graphite layers. The wear rates were however an order of magnitude lower at 2.09 m/s as compared to the wear at 1.09 m/s. The friction coefficients were comparable to those of dry PTFE. In lubricated conditions GP showed a range of friction coefficients similar to lubricated PTFE at the two speeds considered. But the wear rates had a narrow range for both speeds and were much lower than for the lubricated PTFE case. This may be because the influence of increased contact at higher speed is offset by the lower wear rate at the contacts. This observation is based on the fact that dry wear at the high speed condition is an order of magnitude lower than at the lower speed as stated earlier. The lubricated wear rates were comparable to the high speed dry wear of GP at the speed of 2.09 m/s. Wear mechanisms involved were not investigated in this case.

Another interesting example in this area is the work reported by Mens and de Gee [30]. They studied the friction and wear properties of several polymer blends, which in some cases included glass fibres also. The tests were conducted in air and water environments. Water may increase or decrease wear rates depending on the polymer. For example the wear rate of PEEK was 14.8×10^{-15} in the dry condition while it reduced to 1.4×10^{-15} m^3/Nm in water. On the other hand with polyetherimide the wear rate increased by three times in water. The influence of polymer blending and glass fibres is again different for different polymers. The tests were conducted at a nominal pressure of 5 MPa for a duration of 20 hours at room temperature. The sliding speed was 0.25 m/s. The *f* values ranged from 0.2 to 0.78 for dry conditions while they ranged from 0.08 to 0.40 in water. Thus with some exceptions most of the tests in water did not show significant hydrodynamic effects and friction reduction in water may be attributed to the boundary effects. This work based on long duration tests is of practical relevance when polymers are to be used for water lubricated bearings or in humid atmospheres.

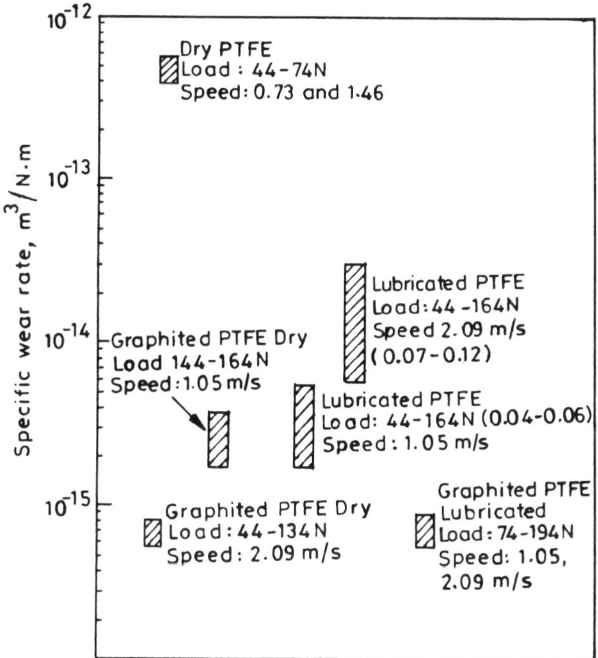

Fig. 6.3. Histogram of wear behaviour on a log scale.
(Reproduced from Ref. [29] by permission of Leaf Coppin Publishing Ltd)

Another aspect of lubricated wear of composites is the issue of filler enrichment on the surface. This has been observed in dry sliding conditions as discussed earlier. When a lubricant is used it can interfere in the transfer process and it is necessary to investigate this problem. A brief study of this aspect has been recently conducted by Choudhary [31]. He studied two bronze and graphite filled PTFE materials with the following composition:

Composite D 25% bronze + 5% graphite + 70% PTFE
Composite G 55% bronze + 5% graphite + 40% PTFE

The wear tests were conducted in a 2 pin-on-disk machine in dry and lubricated conditions. The tests were conducted for a duration of 90 minutes at a sliding speed of 0.71 m/s. The comparative wear data obtained are shown in Table 6.1 that includes the Cu/F ratio on the pin surface. This ratio was based on the relative intensity of Cu and F peaks observed by EPMA over the total pin area. Copper intensity was taken as representative of the bronze content. On the basis of the Cu/F ratio, bronze enrichment was observed on the worn pin surfaces. It is interesting to observe that the enrichment was significantly higher for lubricated contact as compared to the dry contact with one exception for composite D. As may be seen the lubricant decreased the wear rate in all these cases. No investigation was done to understand why enrichment was higher for lubricated cases. Such enrichment is expected to have a positive effect on lubricated wear. Examination of steel counter faces showed transfer films in all cases.

Table 6.1. Specific wear rate and corresponding copper to fluorine ratio for composites D and G. (Reproduced from Ref. [31])

Material	PTFE Composites G				PTFE composite D			
Test condition	Dry		Lubricated		Dry		Lubricated	
Load (N)	Swr (m^3/Nm) x10^{-15}	Cu/F	Swr (m^3/Nm) x10^{-15}	Cu/F	Swr (m^3/Nm) x10^{-15}	Cu/F	Swr (m^3/Nm) x10^{-15}	Cu/F
318	13.70	2.70	4.56	*	54.91	0.66	9.15	0.41
530	28.77	2.44	12.33	5.95	102.50	0.24	25.60	0.96

Swr = Specific wear rate
* Fluorine could not be detected in this case
Cu/F ratio of virgin surface for G ≅ 0.27 and for D < 0.10

6.3.2 Observations on lubricated wear mechanisms

Detailed studies on mechanisms related to liquid lubricants are not available in the literature. As such only some observations are made here on the basis of experimental studies discussed above. This discussion also points to the need for further investigation in specific areas.

Firstly there seems to be no doubt that even in nominally parallel area contacts partial hydrodynamic effects persist as observed above [28,29]. Asperity deformation is likely to be one of the important factors in promoting hydrodynamic effects. There can also be local wedge effects in contact due to non-uniform wear that are more difficult to analyse. Clear theoretical analysis of these aspects is necessary. Such an analysis is available for artificial joints with a metal-UHMWPE combination lubricated with water and other fluids [32]. It can form a useful basis for analysis of hydrodynamic effects. The problem can only be understood by specially designed experiments. Most of the investigations were conducted to study wear and friction, and hydrodynamic effects are at best assessed in a qualitative manner. While low friction coefficients do indicate hydrodynamic effects, detailed analysis should involve estimation of film thickness and the extent of boundary contact through such films.

Polymer wear can be attributed to a degree of polymer-counter face contacts through boundary films. These contacts will include those to the metallic surface as well as to the transferred polymer layers. If lubricant can weaken the polymer adhesion to the counter face the removal rate will be higher. On the other hand, the intervening hydrodynamic and boundary films will reduce transfer to the counter face. It is proposed that the final wear is the result of these competing mechanisms. This aspect is of importance and merits detailed investigation. For example, low wear in a given lubricated system may be only because of reduced transfer though the adhesion to the surface itself may be weaker. Hence there is a need to understand the adhesion strength of polymers besides the hydrodynamic effects. With polymer composites the influence of these effects will be more complex as filler particles are also involved in the boundary lubrication process. They also influence the elastic properties of the near surface material. Hence the analysis may have to bc first confined to pure polymers.

One case has been cited where significant enrichment of bronze was observed when bronze and graphite filled PTFE was slid against steel. This enrichment was higher for lubricated contacts than the dry contacts. This aspect may be significant for lubricated wear of composites and deserves further investigation. Observation of counter faces revealed transfer films, which is also a necessary condition for effective wear control. In some cases visible transfer may not be observed with some fluids as noted in the earlier literature by Evans [33]. His systematic work with several fluids and polymers also showed the importance of solubility parameters in the wear process. Directionally, deleterious effects are higher with those fluids in which the polymer is more soluble. Hence solubility of the polymer in the selected fluid should be taken into account.

The data with regard to strength of adsorption of different fluids on polymers is not available. It is generally expected that on low energy surfaces the adsorption of fluids will be weaker as compared to metals. However the adsorption effects on polymers may be important if these surface become active due to processes like molecular scission. Adsorption effects will be significant for the metallic counter face. With composites containing metallic fillers strong adsorption can occur on their surfaces. No detailed studies are available on these issues. Development of knowledge in this area can be of importance in the selection of fluids. The polymer surfaces can also be modified chemically. One interesting example of such modification is the grafting of acrylic acid groups on polyethylene surfaces [34]. This study dealt with the influence of surface modification on polymer-polymer friction. Such modifications may be of relevance for effective lubrication of polymers if adsorption strength can be increased.

6.3.3 Application aspects

Dry and lubricated polymer bearings are extensively used for light duty applications and are described in the literature [35]. Several such bearings are commercially available with a wide variety of materials and designs. Current interest is in heavy-duty applications. While there has been extensive investigation of materials on laboratory machines for severe conditions, reported information in real systems is limited. Three examples of the reported work are considered here. The first two examples refer to lubricated bearings and the third example considers a high temperature dry bearing application.

Zhang et al [36] have recently demonstrated the advantage of oil lubrication of PTFE based bearings. The bearing consisted of PTFE and Pb composite that is impregnated into a porous bronze layer. The porous bronze layer in turn was anchored to the steel backing through a copper layer. The load carrying capacity of the dry bearing as determined by Chinese standard procedure was 8.82 MPa x m/s. This is the pv (pressure x velocity) value for failure. The same bearing was operated in lubricated conditions. The lubricants studied were engine oil, triethanolamine, and glycerol. When the lubricants were used the load carrying capacity increased by 14 times and the bearings ran cooler and showed much lower wear. Among the three lubricants engine oil showed the lowest wear. Details of the lubrication system were not given but the low friction coefficients with a maximum value of 0.005 clearly indicate strong hydrodynamic effects. Such effects should be mainly due to the usual hydrodynamic action of the designed bearing. Film transfer and lead enrichment on the polymer surfaces were observed. It is not possible to comment on how this transfer was taking place.

The above example is an interesting case for switching to lubrication of polymer bearings. The author stated that since water is known to be deleterious it was necessary to make sure lubricants work satisfactorily. This is a general concern of industry. This aspect can be easily checked in laboratory machines if such information is not already available in literature. Special attention is needed when corrosive or humid environments are involved.

The second example is the replacement of conventional babbit material for thrust pad application. Hydrogenerator thrust pads normally operate at pressures of 3-5 MPa. Applications with higher pressures also exist. When babbit pads are used the friction and wear during start-up is high in the boundary regime. To avoid this problem, the rotor is jacked up with a separate hydrostatic system. This can be overcome if the babbit is replaced by another material with low friction. Pure PTFE pads have been developed and used in Russia but the design details are not clear. They are claimed to operate up to 11 MPa with low friction and wear which is an additional advantage. Recent development in this direction in the U.K. has also been reported [37]. The design consisted of pure PTFE pads joined to steel backing via a bronze wire mesh. Initial trials with these pads have been successful. In all these cases turbine oil was used as the lubricant.

Choudhary et al [38] have adopted a different approach to develop thrust pads. Firstly they have selected bronze + graphite composite pads that have lower wear and similar friction as compared to pure PTFE. Such pads also have higher strength and better heat transfer. On the basis of pin-on-disk tests in dry and lubricated conditions PTFE with 55% bronze and 5% graphite was selected. The turbine oil used was ISO 57 grade with a viscosity of 6 x 10^{-3} Ns/m^2 at 40°C. Simulation tests were then conducted in a scaled down thrust bearing rig. The pads were directly adhered to the steel backing by a selected adhesive. The pad surface was parallel to the runner surface. As this new technique may have problems of debonding, the strength of adhesion was studied in cyclic bending tests. There was no debonding for more than 10^5 cycles with a high shear stress of 9.7 MPa at the interface. This provided adequate confidence in selecting this approach. The thrust bearing tests were conducted for 200 hours with 50 start and stop cycles. The results were compared with those of babbit pads. The performance of the composite pads was superior to babbit both with regard to friction and wear. The applied pressure was 0.32 MPa. Partial hydrodynamic effects occurred even when the rig was operated at the lowest speed of 1.54 m/s. The problem is similar to the laboratory evaluations where hydrodynamic effects persist in nominally parallel contacts as discussed in the previous sub-section. The full details of the development work are given in the cited reference.

The above work again shows the advantages of polymer bearings in lubricated contacts. This is a case where the lubricant is pre-determined. There can be other applications where a choice of lubricants is possible. Basic investigations of the boundary effects and transfer mechanisms will be valuable in these applications.

Another interesting example of bearing development for dry conditions has been reported by Marx and Junghans [39]. They have studied PEEK based materials with a well instrumented dry bearing tester. They found that the bearing material with PEEK + 10% carbon fibre + 10% PTFE + 10% graphite showed the best performance amongst the tested materials. They emphasised the importance of tests in actual bearings. Several aspects like heat transfer cannot be simulated in small scale tests. The selected bearing can operate effectively up to 230°C. This example shows the potential of specialised polymer bearings. It may be observed that some situations demand application under dry conditions where lubrication circuitry is avoided. But there can be many situations where a switch over to lubricated bearings will be of advantage.

6.4. Dry wear of ceramics

The manufacture of ceramics involves basically two steps that involve green shaping and densification. The raw ceramic in powder form is first shaped by pressing, extrusion, or other methods. The densification is done by hot pressing, isostatic pressing, or other specialised processes like reaction bonding. Sintering aids are added where required to control the microstructure and improve the toughness. With the complex processing involved it is not easy to control the structure and porosity of ceramics. Also different processing routes will result in different properties. Ceramic materials used in tribological applications have been introduced in section 1.5.1. Wear mechanisms and modelling are first considered. The present approach based on wear maps is then presented. The properties of typical ceramics are included in this coverage. It is considered that in a limited coverage this is the best way to obtain an overview of the dry wear.

6.4.1 Wear mechanisms and modelling

Wear mechanisms involved with ceramics are complex. The mechanisms involve tribochemical interactions, abrasion, microfracture, large scale fracture, and fatigue. Additional problems include adhering transfer layers whose influence on wear is ill defined. The detailed equations are not considered in this section but they may have partial success in modelling within a narrow range of conditions.

The present consensus seems to emphasise the wear map approach and the development of semi-empirical equations for wear modelling. Wear map approach will be considered in the next sub-section.

One dominant factor in ceramic wear is the asperity level fracture. The maximum tensile stress at the asperity will be the governing stress. This value, σ_{max}, is $k'p_0$ in which k' is evaluated from Eq. 3.6. Stress intensity factor K_I has been defined in section 3.6.1 and is

$$K_I = B\sigma(\pi a_c)^{0.5} \tag{6.1a}$$

From this equation it can be shown that severe wear occurs when

$$B\sigma_{max}\sqrt{\pi a_c} \geq K_{IC} \tag{6.1b}$$

where B is the geometric factor, a_c is the typical crack dimension, σ_{max} is the maximum tensile stress, and K_{IC} is the fracture toughness expressed in $Nm^{-3/2}$. In other words it is assumed that fracture leading to wear occurs when the stress intensity factor exceeds the fracture toughness of the material. Expressing k' in terms of Eqn. 3.6 and assuming typical ν value of 0.25 for ceramics it can be shown for a circular contact

$$\sigma_{max} = p_0(1+10f)/6 \tag{6.2}$$

where p_0 is the maximum Hertzian contact stress at the asperity.

Combining the Eqs. 6.1b and 6.2 it can be shown that for mild wear

$$(1+10f)p_0\sqrt{a_c}\,/\,K_{IC} \leq 6/(B\sqrt{\pi}) \tag{6.3}$$

The right hand side of the equation is a constant designated as C_m. The left hand side is referred to as the severity index S_{cm}. This concept and its modifications have been extensively used in modelling ceramic wear. Firstly it is necessary to define a_c. Adachi [40] and Hsu and Shen [41] have taken mean grain size as the typical value. The modification to the S_{cm} involves thermal effects, which can intensify crack propagation due to thermal shock. Such effects are evaluated on the basis of contact temperature rise.

While the transition to severe wear can be defined as above this does not provide a means of modelling wear as a function of severity index. It is however reasonable to expect that the wear will increase with severity index. The wear volume is considered to be proportional to the real contact area that may be estimated from Eq. 1.7. Another factor to be considered is the possibility of fracture at the level of geometric contact. Such a situation can arise when stress levels are high enough to promote fracture in the gross contact area. Under such circumstances the crack length is taken to be the radius of geometric contact. Such wear has been referred to as 'ultra severe wear' in [41].

6.4.2 Ceramic wear maps

Definitive equations to predict ceramic wear over a wide range of operating conditions are not available. The current approach is to develop wear maps, which firstly relate wear rates to operating conditions. The next step is to model the wear rates in terms of semi-empirical equations centred round the severity index. A detailed account of the wear map approach to ceramic wear has been recently published by Hsu and Shen [41]. These authors studied alumina, Yttrium doped zirconia, silicon carbide, and silicon nitride over a wide range of operating conditions in dry as well as lubricated conditions. Table 6.2 provides information on the properties of the ceramics used. In this table Y-TZP refers to the yttrium doped tetragonal zirconia. The large variation between different ceramics with regard to toughness and thermal properties may be noted. The value of the severity index, and the associated wear are strongly influenced by these properties. The authors used a modified 4-Ball configuration in which the ball slid on three flat surfaces. The load range was 2 to 360 N while the speeds ranged from 1.9 to 570 mm/s. The load influence was studied by step-wise loading at a given speed. At each step the test was run for 5 minutes. Thus wear evaluation was based on very short duration tests. The lubricants studied were purified paraffin oil as well as water. Three dimensional wear maps were constructed for all conditions as a function of load and sliding speed. The various wear mechanisms observed in different zones were also identified. These wear mechanism maps for dry conditions are reproduced in Fig. 6.4. They have identified wear modes that include micro-abrasion and different forms of fracture. There are also major differences in the response of each material to stress and sliding speed. Such maps provide useful guidance with regard to the zones of safe operation.

The wear rates mentioned here are in mm^3/s. It will be useful to relate the wear rate to known parameters. The mild, severe, and ultra severe wear can be expressed by three different equations. It is more convenient to model the wear in all modes by a single equation. The search for correlation under dry conditions indicated that wear volume can be reasonably correlated to

$$\sigma_{max}(T^*/T_0)Nl$$

where T^* is the temperature rise in the nominal contact area, T_0 is ambient temperature, N is the normal load, and l is the sliding distance. The ratio of the two temperatures is considered to be directionally related to the influence of thermal shock on fracture.

It may be noted that the studies were confined to ceramics sliding on themselves. The efforts in this direction are continuing. The authors also discussed the possibilities and limitations of such approaches. The issues involved include long-term effects like fatigue and possible tribochemical interactions even under nominally dry conditions. Also the question whether wear rates observed in step-load tests are representative of the steady state need to be addressed.

Table 6.2. Material properties of the ceramics in wear map studies. (Reproduced from Ref. [41] by permission of CRC Press)

Description	Al_2O_3	Y-TZP	SiC	Si_3N_4
Process	Sintered	Sintered	Sintered and post-HIP	HIP[a]
Sintering aid/Impurities	—	Al_2O_3, Y_2O_3	Al	Mg, Fe, Al, W
Density, g/cm^3	3.9	6.05	>3.17	3.25
Phase		t-ZrO_2		
Average grain size, m	2–15	~1.0	3–8	0.3–2
Elastic modulus, GPa	372	220	430	310
Poisson's ratio	0.22	0.28	0.16	0.28
Hardness, GPa (20°C)	16 ± 0.8	13 ± 0.7	31 ± 1.6	24 ± 1.2
(500°C)	8 ± 0.4	4 ± 0.2	18 ± 0.9	17 ± 0.9
(1000°C)	4 ± 0.2	3 ± 0.2	10 ± 0.5	13 ± 0.7
Fracture toughness, MPa.m$^{1/2}$	4.5	8.5	3.2	5.4
Compressive strength, GPa	2.6	1.9	2.5	3.0
Specific heat, J/g°C	0.88	0.4	0.95	0.65
Thermal conductivity, W/m°C	35.6	1.8	110	33.0
Thermal expansion, 1/°C	7.1×10^{-6}	1.0×10^{-6}	4.1×10^{-6}	3.5×10^{-6}

[a] HIP = hot-isostatically-pressed

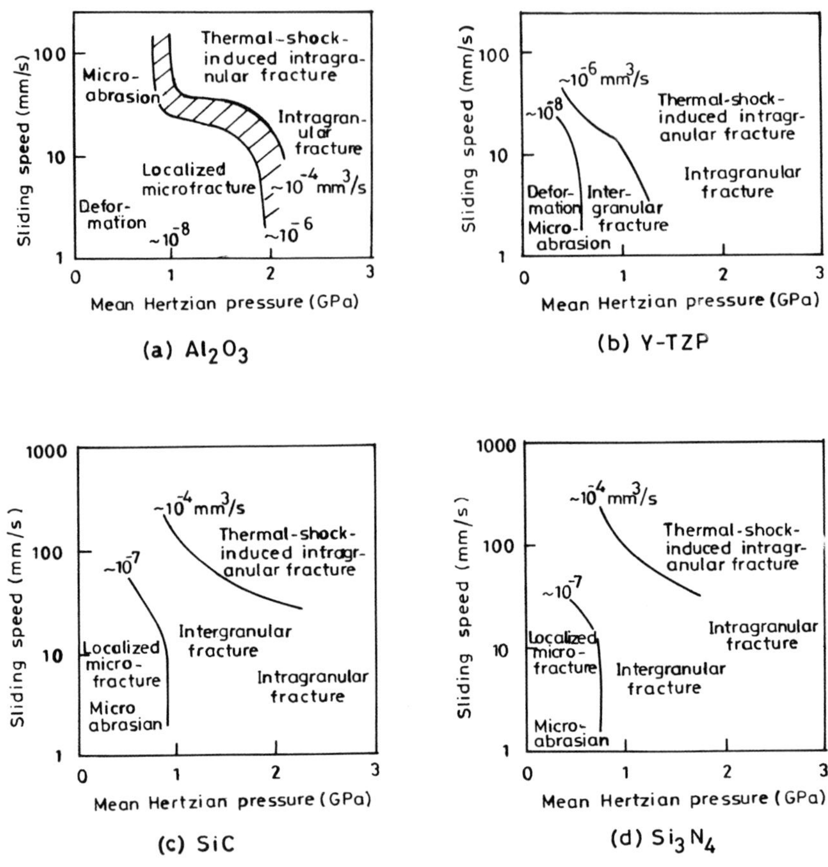

Fig. 6.4. Wear mechanism maps under dry sliding conditions. (Reproduced from Ref. [41] by permission of CRC Press)

Adachi [40] on the other hand attempted to relate wear volume directly to the severity indices with and without thermal effects in dry sliding. The tests were conducted in a pin-on-disk machine with hemispherical pins. Alumina, silicon carbide, and silicon nitride were the ceramics tested. Sliding speeds ranged from 0.01 to 2.3 m/s in most cases. In one case speeds up to 52 m/s were employed. The load range used was 2 to 100 N. Scatter was significant though a rough correlation could be seen. Also the temperature rise calculated in this case was based on the asperity rise while in [41] the overall temperature rise was considered. The specific wear rates in the mild regime can vary between 10^{-9} to 10^{-6} mm^3/Nm. Usually for metals in dry sliding the minimum wear rates observed are nearly 10^{-6} and so there is a scope to use ceramics in dry contacts provided one stays in the low wear zone. On the other hand in severe wear ceramics can show wear rates as high as 10^{-2}.

The empirical correlations only provide a global picture. For example in the mild wear region the relative contributions due to different mechanisms is not known but the wear rate is correlated nevertheless to the empirical parameter selected. Such relation can be specific to the machine and operating conditions selected. In the test configurations used in the above examples as wear occurs, the initial point contact changes to area contact. Thus the overall stress levels decrease with increasing wear. If continuous Hertzian stress acts as in the case of rolling/sliding the nature of the empirical relations may change. Development of large databases may take more time. In the meanwhile the obvious methodology for selection is to collect at least partial data for the operating range involved. The situation is akin to the dry and lubricated wear of metallic materials discussed in earlier chapters. An additional complication involved is the sensitivity of ceramics to wear. Even minor changes can affect the wear response and this creates difficulties in generalising data. Several other papers on ceramic wear maps are available in the literature with varying nomenclature [42,43,44]. In the present section a consideration is given only to the two recent and comprehensive approaches to wear maps. It is considered that this is adequate to meet the objective of this chapter. More recently attempts have also been made to construct ceramic-metal wear maps [45].

6.5 Lubricated wear of ceramics

Tribological application of ceramics can involve dry as well as lubricated conditions. For moderate temperature applications fluid lubricants with or without additives offer a means of controlling ceramic wear and friction. The lubricant behaviour is specific to the ceramic and cannot be generalised. For high temperature applications fluid lubricants cannot be used and new approaches are necessary. Moderate applications can involve ceramic-ceramic as well as metal-ceramic combinations. In the present section lubricant-ceramic interactions at moderate temperatures are first considered. This is followed by applications at moderate and high temperatures. The word 'moderate' usually represents applications not exceeding 250°C. Beyond 150°C synthetic fluids have to be used for lubrication. The types of synthetic fluids have been covered in chapter 2.

6.5.1 Ceramic-lubricant interactions

In general terms [46] the adsorption of hydrocarbons is weak on ceramic surfaces. With polar molecules like fatty acids the adsorption is strong on oxide ceramics. This is similar to the situation with metallic oxides. On the other hand their adsorption is weak on non-oxide ceramics like SiC. Water has a significant influence on the tribochemical reactions. It is known to interact with most of the ceramics resulting in the formation of oxides and hydroxides of variable composition. As an example alumina can form different types of hydroxides on the surface [47]. In the case of silicon nitride significant surface interaction with the formation of hydrous silicon oxides is observed. This can also lead to surface smoothening resulting in hydrodynamic effects [48]. The effect of such chemical interactions on wear is variable and depends on the operating conditions. The detailed wear maps generated in presence of water and cited earlier [41] showed, that the influence of water is variable and depends on the tribochemical reactions as well as its influence on crack growth. In this connection an important review of work till the end of the eighties by Lancaster [49] is of importance. He observed that besides tribochemical reaction, water could affect crack growth. It can also modify the aggregation of the wear debris. The aggregated wear debris in some cases can act as a wear resistant third body. Disruption of such layers can increase the wear rate. The overall influence will be a combination of the three effects involved.

An interesting attempt has been made to distinguish the relative proportion of chemical and mechanical wear with silicon nitride sliding on itself in water [50]. The chemical reaction generates ammonia, which is soluble in water. One mole of the nitride results in four moles of NH_3. By measuring the ammonia concentration the equivalent chemical wear can be calculated. The total wear can then be separated into 'chemical' and 'mechanical'. When wear rates were low the proportion of chemical wear was more significant while mechanical wear predominated when wear rates were high. This idea can be put to practical use. Many industrial lubricants will contain dissolved water. Determination of the ratio of total silicon to that present as nitride in the wear particles can be a useful guide to assess the extent of tribochemical interaction with water. In real systems attention has to be paid to the possible ingress of silica particles, which can affect the ratio.

One point of interest here is to know how the 'mechanical' wear is influenced by water. This influence is complex and cannot be predicted theoretically. With regard to humidity one possible approach can be to first obtain steady wear situation in dry air. Then humidity may be introduced at the required level and the change in wear rate can be obtained. The changes can be reliable only when they differ adequately from the mean value for the dry case. One interesting example [51] of such an approach from another area of tribology may be mentioned. To study the influence of friction modifiers steady friction was first obtained. The friction modifier was then introduced into the system and the change in friction noted. The usual procedure of evaluation with and without friction modifier involves a large number of tests, as statistical variations have to be accounted for. The author considers this concept of differential change should be more actively explored in tribology.

The role of base fluids on ceramic tribology has also been studied though less extensively than with water. Hydrocarbon fluids in many cases were found effective in reducing friction and wear of ceramics. This improvement is related to the chemical interaction with the surfaces though the adsorption is expected to be weak. One interesting example is the study on friction and wear of silicon nitride sliding on itself by Jahanmir et al [52]. The study was conducted with pins of hemispherical cap of 3.0 mm radius. The tests were conducted at a low sliding speed of 1.0 mm/s at a load of 9.8 N. Clear evidence of tribochemical reactions was observed with water as well as humid air as significant amorphous silica was detected in the wear track. The tests with hexadecane showed an order of magnitude reduction in wear compared to water. There was clear evidence of carbon formation as the reaction product on the wear track showed amorphous silica as well as carbon. The reaction to form silica was attributed to dissolved water present in the hydrocarbon. A satisfactory explanation for chemical interaction of the hydrocarbon was not available. The authors considered decomposition is possibly due to the catalytic activity of the ceramic. This work is also a pointer to the significant role that dissolved water can play in tribochemical reactions.

Klaus et al 53] studied the lubricated wear of silicon nitride with various lubricants in a 4- Ball machine in the machine load range of 10-40 kg and a speed of 600 rpm. Steel-steel, steel-ceramic and ceramic-ceramic combinations were studied. The lower balls were replaced by ceramic flats. It is of interest to note that a steel-ceramic combination was vastly superior to a steel-steel system with the lubricants studied. On the other hand a ceramic-ceramic couple showed comparable performance to a steel-ceramic couple in some cases while in other cases it showed catastrophic wear. The authors directionally discussed the efficacy in terms of friction polymer formation at the high temperatures involved. The lubricants studied included mineral oil, polyglycol, polyol ester, engine oil, and tricresyl phosphate (TCP) and tributyl phosphate. They also observed superior performance with 1% TCP in mineral oil as compared to neat TCP. The comparisons were based on short duration tests of 30 minutes. The synergistic effect of a metal-ceramic couple is interesting and deserves further investigation.

The detailed information for lubrication with purified paraffin oil for four different ceramics has been plotted in the earlier cited reference [41]. The paraffin oil that is a representative hydrocarbon lubricant was found beneficial for all the four ceramics studied. In all cases paraffin was found to be useful in reducing wear. In particular the speed effects were significantly lower as compared to the dry case. This was attributed to the low interfacial temperatures in the presence of lubricant. The liquid paraffin was considered to be inert and hence the possible chemical reactions were not taken into account. On the other hand Jahanmir et al [52] found carbon formation with hexadecane at very moderate sliding conditions as discussed earlier. As the conditions in the wear map studies were more severe, carbon formation should be expected in this case also with liquid paraffin. Information in this regard is not available and cannot be commented upon. Detailed wear maps with other fluids for different ceramics is not available. It is reasonable to expect a positive influence with lubricants over a wide range.

The above studies are based on conventional lubricants and additives. These are all well studied with metallic contacts and a similar base may eventually develop for ceramics. But the question is whether there can be an all-together different chemistry suitable for ceramics. This is likely to be addressed in the future.

6.5.2 Moderate temperature applications

The practical application of ceramics normally involves replacement of existing components with ceramic-ceramic or metal-ceramic combinations. The metal-ceramic combination is also referred to as a hybrid combination in the literature. In many situations the lubricant is pre-decided due to the other components in the system. In some cases a new formulation is possible if the component is exclusively lubricated. For example one can envisage a new formulation for a ceramic wire-drawing die. On the other hand if components like piston rings or the cam follower in an engine are to be replaced, they have to function with the existing lubricant that has to satisfy the requirements of the total system. The problem is demanding and is considered below with an example.

Winn et al [54,55] reported on the lubricated wear of ceramics. They studied steel-ceramic combinations in a 3 pin-on-disk machine. Domed pins were used and the initial Hertzian stress applied on each pin was 596 MPa. The tests were conducted at different speeds of 0.06, 0.25, and 1.0 m/s. One lubricant used was the mineral based fully formulated SAE 30 grade engine oil. The second lubricant studied was a synthetic lubricant with a similar additive package. Alumina and silicon nitride were investigated in the tests. Initial work indicated alumina had relatively higher wear under these conditions and the effort was focussed on the nitride ceramic. Tests with durations as long as six months were conducted to study long term wear. This is a unique feature of these studies. Nominal temperature of the pins was maintained at 100°C. The wear rates ranged from 4 x 10^{-12} to 10^{-11} mm^3/Nm and were lower than for the steel-steel system. Surface polishing due to tribochemical reactions was put in evidence. Surface studies confirmed dithiophosphate based films. Partial hydrodynamic effects were observed at the high speed of 1.0 m/s. With ester based lubricants higher initial wear was observed followed by a lower steady state wear. These studies were with reference to ceramic pins sliding on a steel disk. Some tests were also conducted with steel pin sliding against ceramic, which behaved somewhat differently. A possible explanation was attempted. The pin wear volume was obtained neglecting elastic recovery. This study provided adequate confidence for some engine applications like cam followers. If the tests were conducted for short duration of an hour or so, as is the usual practice, the conclusions could be different and less reliable. The study clearly shows that running-in can persist over several kilometres. Earlier observations made by the author in section 5.4.1 with regard to running-in are pertinent in this regard. Similar studies may be necessary to evaluate possible ceramic replacements for valve seats, piston rings, and liners.

One comment regarding wear maps is important from a practical point of view. The present approaches define mild wear as less than 10^{-7} mm^3/Nm and concentrate on the speed-load transitions to severe wear. In lubricated wear, as seen above, the interest is in very low wear rates which are orders of magnitude lower than this value. Also the ability to distinguish small differences in wear rates is important. There is a clear need for the generation of wear maps that provide expanded information in the mild wear region. Development of semi-empirical equations to relate wear rates to operating parameters will also be valuable. The situation is similar to the case of antiwear additives discussed in the previous chapter.

Another area where extensive investigations are being carried out is the development of ceramic rolling element bearings. Silicon nitride is found useful for this purpose. The development of quality ceramics has undergone positive changes and it is now possible to have materials with assured performance. This is unlike the case two decades ago when batch-to-batch variations were common. The bearings have the advantage of high stiffness, lower weight, and high temperature capability. Extensive fatigue tests continue to be conducted with different configurations [56]. Direct bearing tests are also being undertaken. Most of the studies were conducted with lubricants and showed that their performance is better than metallic bearings in many cases. They are already being used in some applications. Future applications are expected for gas turbine bearings with shaft speeds up to 75,000 rpm.

Yet another area in which ceramics are being increasingly used is the artificial human joint. Ceramic-ceramic as well as metal-ceramic combinations are being used in such applications. Alumina is the main ceramic for these applications. Such developments were possible by controlling roughness to nano levels and ensuring significant hydrodynamic effects in the contact [26,32].

6.5.3 High temperature applications

The use of ceramics without liquid lubricants at high temperatures is a desirable goal. If such a goal is achieved, engines for example, can be operated without lubricant. Engine cooling can be eliminated, increasing the thermal efficiency significantly. Another advantage is lower mass, which again contributes to higher efficiency. The major problems to be overcome are the friction and wear, both of which are high for the current ceramics at high temperatures.

Attempts to lubricate ceramics at temperatures beyond 800°C were based on the use of solid lubricants like graphite [57] incorporated in the ceramics. The more recent effort has concentrated on the idea of developing lubricious oxides on the surface. Some metals can form homologous series of non-stoichiometric oxides that can have desirable friction and/or wear properties. The metals with such properties are Ti, V, Mo, and W. Several ceramic composites with TiC and TiN have been studied for high temperature applications. They show an advantage with regard to wear as compared to ceramics alone, due to the formation of several titanium oxides on the surfaces. Inclusion of Mo in the composite further enhances the wear reduction. The mechanisms involved are very complex and detailed surface studies are being conducted as reported by Woydt et al [58]. This paper also provides a good insight into the concept of lubricious oxides.

Two other approaches for high temperature lubrication are being developed. Lauer et al [59] have used the idea of carbon lubrication. The idea is based on the catalytic dehydrogenation of ethylene by a nickel catalyst. They have used several nickel containing composites that included metals and ceramics and demonstrated low friction at 500°C due to carbon film formation in ethylene gas. The second approach is vapour phase lubrication. In this approach additive is introduced in the vapour phase with controlled surface reaction [60]. Tricresyl phosphate was tried for such lubrication with good results. The future of the ideas mentioned here is not yet clear.

6.5.4 Observations on lubricated wear of ceramics

Ceramics are prone to tribochemical reactions, and the influence of water vapour and lubricants has been discussed above. The complexity involved in film formation is similar to the case of metallic contacts. The difference is that metal-metal systems were studied over a longer period of time and the available experience has established the practice. Even here the problems persist when the operating severity is increased. Also the evaluation methods are semi-empirical and there is a need for a more scientific approach. These issues were discussed at some length in the previous chapter. Systematic development of wear maps was proposed to improve the present situation. Similar considerations apply to ceramics intended for application in moderate conditions. Consideration of possible mechanisms and empirical relations should follow from such maps.

The issue of steady state wear is again of importance. From the information in the previous sections wear determinations were based on 5-minute tests in one case. In another case studies of several months were involved to establish the steady state. As these issues were already discussed in detail in the previous chapter they are not further considered here.

Fundamental studies are conducted more from the point of view of developing understanding at the conditions chosen by the researcher. The interest is normally confined to a model that describes the interactions and development of a possible wear equation. Scientific curiosity demands such an approach. This results in an excellent ability to understand mechanisms in a narrow range. Due to the large number of variables in real systems, application of the vast amount of available literature is limited. The need of the hour is to shift focus to the gaps of knowledge in practice. The previous chapter addressed this problem in section 5.3. Similar considerations apply to ceramics. Ceramics are more prone to local fatigue and fracture. The influence of base fluids and additives on crack growth will be an additional consideration in developing basic understanding. Such studies however are useful mainly in moderate temperature applications. With regard to the high temperature applications no suggestions can be made as the available knowledge is inadequate.

References

1. C. M. Pooley and D. Tabor, Friction and molecular structure: The behaviour of some thermoplastics, Proc. Roy. Soc. Series A, 329 (1972) 251-274.
2. K. Tanaka and Y. Uchiyama, Friction, wear, and surface melting of crystalline polymers, Proc. Wear of Materials, ASME, 1977, 499-530.
3. V. A. Smurugov, A. I. Senatrev, V. G. Savkin, V. V. Biran and A. I. Sviridyonok, On PTFE transfer and thermoactivation mechanism of wear, Wear, 158 (1992) 61.
4. S. K. Biswas and K. Vijayan, Friction and wear of PTFE - a review, Wear, 158 (1992) 193.
5. J. K. Atkinson, K. J. Brown, and D. Dowson, The wear of high molecular weight polyethylene, Part I: The wear of isotropic polyethylene against dry steel in unidirectional motion, J. Lub. Tech.. ASME, 100 (1978) 208.
6. V. R. Agarwal, U. T. S. Pillai and A. Sethuramiah, New observations on PTFE wear mechanism, Proc. Wear of Materials, ASME, 1989, 501.
7. K. Marcus and C. Allen, The sliding wear of ultrahigh molecular weight polyethylene in an aqueous environment, Wear, 178 (1994) 17..
8. V. K. Jain and S. Bahadur, Development of a wear equation for polymer-metal sliding in terms of the fatigue and topography of the sliding surfaces, Wear, 60 (1980) 237.
9. N. P. Suh, Tribo Physics, Prentice Hall, New Jersey, 1986, chapter 6.
10. N. Viswanath and D. G. Bellow, Development of an equation for the wear of polymers, Wear, 181-183 (1995) 42.
11. H. Czichos, Influence of adhesive and abrasive mechanisms on the tribological behaviour of thermoplastic polymers, Wear, 88 (1983) 27.
12. W. Hirst and A. E. Hollander, Surface finish and damage in sliding contact, Proc. Roy. Soc. A. 337 (1974) 379.
13. H. Bohm, S. Betz, and A. Ball, The wear resistance of polymers, Trib. Int., 23 (1990) 399.
14. T. A. Stolarski, Tribology of polyetheretherketone, Wear, 158 (1992) 71.
15. D. Konur and F. L. Matthews, Effect of the properties of the constituents on the fatigue performance of composites: a review, Composites, 20 (4), (1989) 317.
16. A. K. El-Senussi and J. P. H. Webber, Critical strain energy release rate during delamination of carbon fibre reinforced plastic laminates, Composites, 20 (3), (1989) 245.
17. K. Friedrich, Z. Lu, and A. M. Hager, Overview of polymer composites for friction and wear application, Theor. Appl. Fracture Mechanics, 19 (1993) 1.
18. K. Friedrich, Z. Lu and A. M. Hager, Recent advances in polymer composites' tribology, Wear, 190 (1995) 139.
19. T. A. Blanchet and F. E. Kennedy, Sliding wear mechanism of polytetrafluoroethylene (PTFE) and PTFE composites, Wear, 153 (1992) 229.

20. S. Bahadur and D. Gong, The transfer and wear of nylon and CuS-nylon composites: filler proportion and counter face characteristics, Wear, 162-164 (1993) 394.
21. S. Bahadur and A. Kapoor, The effect of ZnF_2, ZnS, and PbS fillers on the tribological behaviour of nylon 11, Wear, 155 (1992) 49.
22. Q. Zhao and S. Bahadur, The mechanism of filler action and criterion of filler selection for reducing wear, Wear, 225-229 (1999) 660.
23. K. A. Grosch, The relation between the friction and visco-elastic properties of rubber, Proc. Roy. Soc., London, Series A, 274 (1963) 21.
24. K. Holmberg and G. Wickstrom, Friction and wear tests of polymers, Wear, 115 (1987) 95.
25. A. Unswoth, Recent developments in the tribology of artificial joints, Trib. Int., 28 (1995) 485.
26. D. Dowson, New joints for the millennium: wear control in total replacement hip joints, Proc. Instn. Mech. Engrs., Part H, Journal of Engineering in Medicine, 215 (2001) 335.
27. R. W. Bramham, R. B. King, and J. K. Lancaster, The wear of PTFE-Containing dry bearing liners contaminated by fluids, ASLE Trans., 24 (1981) 479.
28. P. M. Dickens, J. L. Sullivan, and J. K. Lancaster, Speed effects on the dry and lubricated wear of polymers, Wear, 12 (1986) 273.
29. A. Sethuramiah, K. L. Awasthy, Braham Prakash, and P. K. Mahapatra, Lubricated wear of PTFE and graphited PTFE, Lub. Sci., 3 (1991) 181.
30. J. M. M. Mens and A. W. J. de Gee, Friction and wear behaviour of 18 polymers in contact with steel in environments of air and water, Proc. Wear of Materials, ASME, (1991) 563.
31. T. R. Choudhary, Tribological investigation of heavy duty polymer based thrust pads, PhD thesis, IIT Delhi, 2001.
32. D. Jalali-Vahid, M. Jagatia, Z. M. Jin, and D. Dowson, Prediction of lubricating film thickness in a ball-in-socket model with a soft lining representing human natural and artificial joints, Proc. Instn Mech. Engrs., Part H, Journal of Engineering in Medicine, 215 (2001) 363.
33. D. C. Evans, Polymer-fluid interactions in relation to wear, in D. Dowson, M. Godet, and C. M. Taylor (eds.), Proc. 3rd Leeds-Lyon Symposium on Tribology, Instn Mech. Engrs. Publication, 1978, 47-59.
34. L. Lavielle, Polymer-polymer friction: relation to adhesion, Wear, 151 (1991) 63.
35. V. A. Bely, A. I. Sviridenok, M. I. Petrokovets, and V. A. Savkin, Friction and Wear in Polymer-Based Materials, (Translated from the Russian), Pergamon Press, 1982, chapter 9.
36. Z. Zhang, W. Shen, W. Liu, Q. Xue, and T. Li, Tribological properties of polytetrafluoroethylene based composite in different lubricant media, Wear, 196 (1996) 164.

37. J. E. L. Simmons, R. T. Knox, and W. O. Moss, The development of PTFE (polytetrafluoroethylene)-faced hydrodynamic thrust bearings for hydrogenerator application in United Kingdom, Proc. Instn Mech. Engrs., Part J, Journal of Engineering Tribology, 212 (1998) 345.
38. T. R. Choudhary, A. Sethuramiah, O. Prakash, and G. V. Rao, Development of polytetrafluoroethylene composite lining for a hydrogenerator thrust pad application, Proc. Instn Mech. Engrs., Part J, Journal of Engineering Tribology, 214 (2000) 375.
39. S. Marx and R. Junghans, Friction and wear of highly stressed thermoplastic bearings under dry sliding conditions, Wear, 193 (1996) 253.
40. K. Adachi, K. Kato, and N. Chen, Wear map of ceramics, Wear, 203-204, (1997) 291.
41. S. M. Hsu and M. C. Shen, Wear Maps, in Bharat Bhushan (ed.), Modern Tribology Handbook, Vol 1, CRC Press, New York, 2001, chapter 9.
42. K. Hokkirigawa, Wear map of ceramics, Wear of Matrials, ASME, 1991, 353.
43. Y. S. Wang, S. M. Hsu, and R. G. Munro, Ceramic wear maps: alumina, Lub. Eng., (1991) 63.
44. A. Blomberg, M. Olsson, and S. Hogmark, Wear mechanisms and tribo mapping of Al_2O_3 and SiC in dry sliding, Wear, 171 (1994) 77.
45. J. R. Gomes, A. S. Miranda, J. M. Vieira, and R. F. Silva, Sliding speed-temperature transition maps for Si_3N_4/iron alloy couples, Wear, 250 (2001) 293.
46. P. Studt, Boundary lubrication: adsorption of oil additives on steel and ceramic surfaces and its influence on friction and wear, Trib. Int., 22 (1989) 111.
47. R. S. Gates, E. E. Klaus, and S. M. Hsu, Tribochemical mechanism of alumina with water, Trib. Trans., STLE, 32 (1989), 357.
48. T. Saito, Y. Imada, and F. Honda, An analytical observation of the tribochemical reaction of silicon nitride sliding with low friction in aqueous solutions, Wear 205 (1997) 153.
49. J. K. Lancaster, A review of the influence of environmental humidity and water on friction, lubrication, and wear, Trib. Int., 23 (1990) 371.
50. T. Saito, T. Hosoe, and F. Honda, Chemical wear of sintered Si_3N_4, hBN and Si_3N_4-hBN composites by water lubrication, Wear, 247 (2001) 223.
51. S. G. Arabyan, I. A. Holomonov, A. K. Karaulov, and A. B. Vipper, An investigation of the effectiveness of antifriction additives in motor oils by laboratory methods and engine tests, Lub. Sci., 5 (1993) 241.
52. S. Jahanmir and T. E. Fischer, Friction and wear of silicon nitride lubricated in humid air, water, hexadecane, and hexadecane + 0.5% stearic acid, Trib. Trans., STLE, 31 (1987) 31.
53. E. E. Klaus, J. L. Duda. and W. T. Wu, Lubricated wear of silicon nitride, Lub. Eng., 47 (1990) 679.

54. A. J. Winn, D. Dowson, and J. C. Bell, The lubricated wear of ceramics, Part 1: The wear and friction of silicon nitride, alumina and steel in the presence of a mineral oil based lubricant, Trib.Int., 28 (1995) 383.
55. A. J. Winn, D. Dowson, and J. C. Bell, The lubricated wear of ceramics, Part I: The wear and friction of silicon nitride, alumina and steel in the presence of a mineral oil based lubricant, Trib.Int., 28 (1995) 383.
56. M. Hadfield and T. A. Stolarski, The effect of the test machine on the failure mode in lubricated rolling contact of silicon nitride, Trib. Int., 28 (1995) 377.
57. A. Gangopadhyay and S. Jahanmir, Friction and wear characteristics of silicon nitride-graphite and alumina-graphite composites, Trib. Trans., STLE, 34 (1991) 257.
58. M. Woydt, A. Skopp, I. Dorfel, and K. Witke, Wear engineering oxides/antiwear oxides, Trib. Trans., STLE, 42 (1999) 21
59. J. L. Lauer and B. G. Bunting, High temperature solid lubrication by catalytically generated carbon, Trib. Trans, STLE, 31 (1987) 339.
60. E. E. Gramham and E. E. Klaus, Lubrication from vapour phase at high temperature, ASLE Trans., 29 (1986) 229.

Nomenclature

a_c	typical crack dimension
B	geometric factor
C_m	constant= $6/(B\sqrt{\pi})$
f	coefficient of friction
k'	multiplication factor to obtain σ_{max}
K_I	stress intensity factor
K_{IC}	fracture toughness
l	sliding distance
N	normal load
p_0	maximum Hertzian contact stress at the asperity
S_{cm}	severity index
T^*	temperature rise in the nominal contact area
T_0	ambient temperature

Greek Letters

ν	Poisson ratio
σ_{max}	maximum tensile stress

7. Tribological evaluation methodologies

7.1 Introduction

The role of lubricants in controlling wear and scuffing has been discussed in chapter 5. As discussed it is not possible to predict performance from fundamental considerations alone. The ideal approach is to evaluate lubricants in real systems. Such direct evaluations are done in some cases. Prescribed engine tests to evaluate wear and fuel efficiency is an example of such an approach. Such testing is expensive and time consuming. Also industrial lubricants are used in a variety of applications and operating conditions and it is impractical to use real systems for evaluation. Experimental evaluation in tribological rigs is the present available approach to assess performance. Several standard rigs from ASTM, IP, DIN, and other bodies are available for testing. In addition several non-standard in-house techniques are also used that are aimed at improving the evaluation capability. The performance level required is specified for a given lubricant. All such methods have their possibilities and limitations. Critical evaluation of the test methods is attempted in this chapter with the background available from chapter 5. All the available test methods will not be covered. The emphasis will be on the methodology centred round well known tests.

At the outset it is necessary to realise that performance considerations are limited here to lubricant-metal interactions. In some cases the performance problems are related to other factors like ingress of abrasives, starvation of the lubricant, and quality control of the materials. Careful analysis of a tribological problem is necessary to assess whether the problem is due to the lubricant or otherwise.

The first section of the chapter considers different contact geometries used in tribological testing and their implications. The second section deals with the evaluation of antiwear additives while the third section deals with EP evaluation. The final section deals briefly with the performance evaluation of metal working lubricants. Issues of boundary lubrication in metal working are also considered in the final section and a new approach is suggested. Other aspects of metal working are covered very briefly. The major objective of this chapter is to develop a more logical framework for tribological evaluation with smaller test rigs.

7.2 Test configurations

The common test geometries used to study wear include pin-on-flat, 4-Ball, ring-on-flat, pin and V-block, and rolling/sliding disk contact. The geometries involved are sketched in Fig. 7.1. Other geometries like crossed cylinders are also used in some cases. Disk machines which run with line contact have an essentially constant Hertzian stress throughout the test for a given normal load. This geometry can simulate gear contact conditions and has been widely used to evaluate scuffing with mineral oils as discussed in chapter 5. The geometry is also used to study fatigue wear extensively. The other machines are normally operated in unidirectional sliding with one surface stationary. In some test machines reciprocating sliding is adopted with one specimen oscillating against a fixed specimen with a selected frequency and stroke length as considered in the later part of the chapter. The 4-Ball machine has a unique tetrahedral geometry. In the pin-on-disk configurations the pin can be flat or hemispherical. In some cases pin-on-cylinder geometry can also be used and the geometry is similar to that of a block-on-cylinder with the block replaced by the smaller pin. The disk can rotate in a horizontal or vertical plane. Debris accumulation is less likely when the disk rotates in a vertical plane.

The testers used for lubricant evaluations are mainly 4-Ball, ring-on-block, and pin and V-Block configurations. Reciprocating testers with varied types of specimens oscillating against a flat are also used. The normal force is applied by lever loading, pneumatic loading or hydraulic loading. The friction force is measured by transducers of different kinds that respond to the frictional force acting on the stationary specimen. Commercial machines adopted in standards are well designed ensuring uniform normal loading and accurate measurement of frictional force. The stresses acting in line contact at the start of the test can be calculated on the basis of equations given in section 5.2. In the case of the 4-Ball machine the load acting on one ball will be 0.408 L where L is the machine load. The steel balls are of 12.7 mm. The initial elastic contact diameter can be expressed in terms of machine load L as follows:

$$d_H = 0.0873(L)^{1/3} \tag{7.1}$$

where d_H is Hertzian diameter in mm and L is machine load in kg

In the case of V-Block the load acting at the contact in terms of machine load L in kg depends on the geometry. In the standard machine as per ASTM D 3233 with a journal radius of 6.35 mm and V-Blocks with 96° angle the load acting at the four line contacts is Lcos42°. The initial stresses at the four line contacts can be obtained from the equations given in section 3.2.

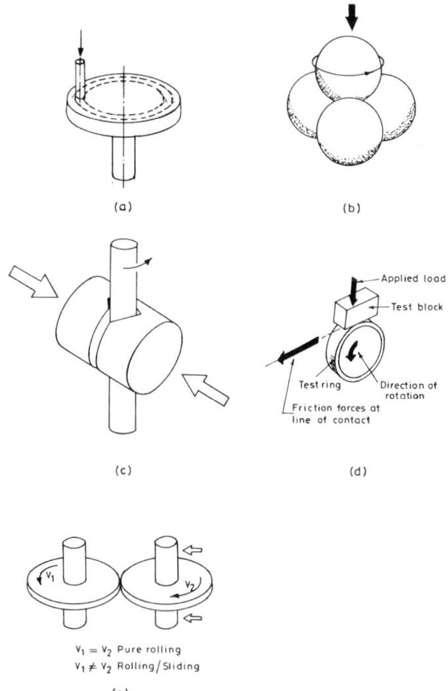

Fig. 7.1. Triobological test machine configurations. (a) Pin-on-disk (b) 4-Ball (c) Pin and V-Block (d) Block-on-ring (e) Disk-disk.

The above machines do not simulate the conditions in the real systems. The possibilities and limitations of such tests will be considered in the later parts of the chapter. Some rigs are also available that completely or partially simulate a sub-assembly. Rating wear in an actual vane pump used in hydraulic systems may be considered a full simulation test. Several available gear tests may be considered partial simulation tests as the gear test is conducted with a specific type of gear combination.

The determination of wear is usually done in these tests by measuring scar dimensions and converting them to wear volume. In many laboratory rigs the wear observed is small and not detectable by weight loss. For example typical wear volumes in a 4-Ball wear test can range between 10^{-3} to 10^{-4} mm^3. The corresponding weight loss is much less than a milligram and cannot be measured. In some cases the scars may not be uniform and approximation by a suitable programme will be necessary. One example of such a situation is the reciprocating tester. The lower stationary surface will have higher wear at the reversal points as compared to the centre.

7.3 Wear evaluation

The evaluation of lubricated wear can in principle be conducted with any test geometry and operating conditions. For the purpose of comparison as well as laying down wear limits standard tests at prescribed conditions are necessary. Several standard tests are available. Development of standards involves a large effort that is reflected in the thoroughness of the test procedures. Test repeatability and reproducibility are also specified. Normally standard tests are conducted at one prescribed set of conditions. Users in turn specify the wear limits based on their requirements. Many test rigs can operate over a wide range of operating conditions and not limited to the prescribed conditions. The purpose of this section is to understand how wear evaluations can be improved. The standard rigs are considered a means to this end. The development of the ideas will be centred round the 4-Ball wear test. This rig is chosen because of its wide spread use. A generalisation is then attempted taking into account other test rigs.

7.3.1 4-Ball rig evaluation

The basic geometry and stresses involved have been considered in the first section. The ASTM D 4172 – 94 describes the standard procedure in detail. The procedure consists of a rotating top ball against the three fixed lower balls immersed in the lubricant. The steel balls are 12.7 mm in diameter and are made of AISI 52100 steel. The oil temperature is controlled by a temperature controller. Pneumatic loading is adapted to obtain good precision. Tests are conducted at a temperature of 75 ± 2°C at a speed of 1200 ± 60 rpm for a duration of 60 ± 1 minutes. The tests are usually conducted at 15 ± 0.2 or 40 ± 0.2 kg load. The average wear scar of the stationary balls forms the basis for wear comparison. The repeatability is specified for these conditions. The tests can also be conducted at any other condition.

Firstly for any comparison of wear behaviour it is necessary to obtain wear rates. For this purpose it is necessary to obtain the wear volume for a given scar diameter and also to obtain the wear coefficient.

The wear volume V of a stationary ball is

$$V = [1.55x10^{-2} d^3 - 1.03x10^{-5} L]d \tag{7.2}$$

where d is the wear scar diameter in mm and L is the machine load in kg.

The above equation takes into account the elastic recovery of the ball. Thus the actual wear volume is less than what is obtained on the basis of final scar only. It is expressed in a convenient form in terms of the scar diameter and machine load. The wear coefficient can be obtained by the following equation

$$K = [VH/23.3(rpm)tx0.408L] \tag{7.3}$$

where H is the hardness in kg/mm^2, and t is the test duration in minutes

The above equations are based on the available literature [1] and are derivable from fundamental considerations [2]. The wear coefficient is based on Archard's approach assuming real area is proportional to hardness. The denominator is the product of sliding distance and the load acting on the ball. The influence of elastic recovery is higher for smaller scar dimensions though this is not readily apparent from the final form of the equation given above. At this stage it is useful to calculate wear volume and wear coefficient based on one hour tests for 40 kg machine load for randomly selected wear scars. Wear coefficient is based on final scar dimension. Hardness is taken as 720 kg/mm^2. The data is given in Table 7.1.

Table 7.1
Wear coefficient calculation for some scar diameters

Wear scar dia, mm	Wear vol, $mm^3 x10^3$	Wear coefficient $x10^8$
0.336	0.0591	0.155
0.44	0.400	1.052
0.54	1.100	2.693
0.61	1.895	4.984

This table shows that wear coefficients can vary by about 2 to 8 times for a variation of the order of 0.1 mm in wear scar. As wear rate expressed in terms of wear volume per unit sliding distance is proportional to the wear coefficient, wear rates have similar variations. These variations are higher when the scar diameters are smaller. The wear rates may also be expressed on the basis of running time using the above equation.

7.3.1.1 Assessment of wear

Firstly it may be observed that as per the standard, repeatability of the wear scar in these tests is 0.12 mm with a 95% confidence limit. Such variations when translated into wear volumes will amount to large variations in wear rates as discussed above. The variations are most likely due to the running–in effects. A typical schematic diagram for two repeat tests 1 and 2 is shown in Fig. 7.2. As wear volume is of importance the plots are made on the basis of cumulative wear volume as a function of time with 15 minute intervals. The lower ball pot is not disturbed and at each stage the wear volume is obtained by average scar diameter and the operation continued until the end of the test. Wear rate in terms of wear volume per unit time considering only the final wear volumes are obtained from the slopes of the D1 and D2 lines. It is evident that significant variation can occur on this basis. If it is assumed that running-in is completed in 30 minutes the slope of the regressed line between 30 and 90 minutes gives the steady state wear rate. The regressed lines are shown as RL1 and RL2 for the two tests. For clarity the regressed line is shown as a part of the overall curve. The wear rates will be lower now as compared to the case based only on the final wear volume. Better repeatability is also expected by such a procedure as the variations due to running-in effects are reduced. The possibilities and limitations of such a procedure will be considered below.

This procedure needs to be assessed from a basic point of view. The running-in and steady state wear have been modelled and empirical equations were developed as given in section 5.4.1. The tests were conducted in a reciprocating tester (RT). The running-in time was taken as the point at which 95% of the wear rate equals the steady state wear rate. Such detailed studies to determine wear rate are the best possible approach with load, temperature, and sliding speed as variables. This will be very useful in mapping wear behaviour in the 4-Ball machine. Programmes developed on this basis will be of utility to the lubricant formulators and users. The importance of such mapping was discussed in chapter 5. One additive can be better or worse than the other depending on the operating conditions. Surely distinction at one or two conditions is inadequate for a formulator to make an effective judgment. Such investigations are possible through cooperative effort. One interesting example of such an effort is the study of transitions in lubrication regimes under different operating conditions by the International Research Group (IRG) on wear [3]. Pending such approaches, at least the wear rate should be based on regression lines as described above. Even here the problem is to know the starting point for regression. Running-in effects will depend on the speed, load, temperature, and the additives in the formulation. Usually visual observation of the curves can indicate the steady state portion of the curve that can be regressed. An example to determine wear rate based on regression of points between 30 and 120 minutes in a 4-Ball machine based on [4] is shown in Fig. 7.3. The figure is based on the tabulated data given in Table 7.2. The tests were conducted at 40 kg and 1200 rpm and 75°C. The purpose of the tests conducted under different conditions was to distinguish between two running-in oils for industry. The wear volume obtained was for separate tests with different time intervals. In each test the variability of the initial wear influences the wear volume. The correlation was hence not very good with correlation coefficients of 0.935 and 0.883 for oils 1 and 2 respectively. The variability observed with oil 1 was lower and hence the regression was better. While better statistical analysis can be done, the inherent variability due to separate tests should be taken into account. This can be eliminated if the wear scar is measured at different time intervals in the same test. Many machines now are equipped to measure wear scar without dismantling the lower pot assembly. With such an approach the regression should give a much better assessment of wear rate. This involves an assumption that the start and stop influences are small. This is a reasonable assumption with conformal wear scars. Comparison of wear rates between additives then becomes more meaningful. Also the total testing time will

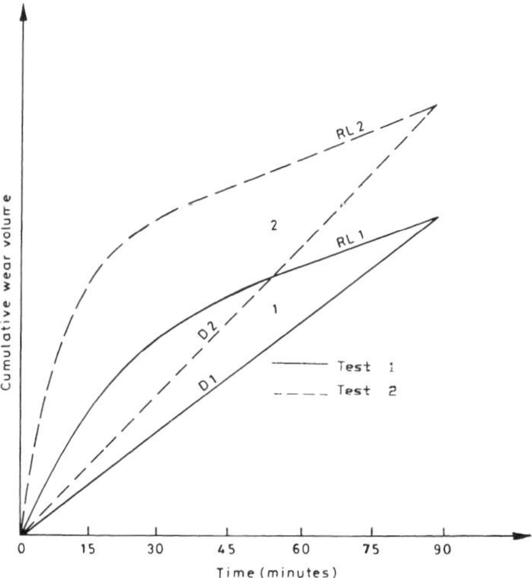

Fig. 7.2. Schematic behaviour of two repeat tests 1 and 2 in 4-Ball tester. RL refers to regression line while D refers to the line drawn from origin to the final wear volume.

Table 7.2
Wear data for two oils in 4-Ball machine

Time (hrs)	Wear volume (oil1) (mm^3) Series 1	Wear volume (oil2) (mm^3) Series 2
0.25	2.11E-04	1.27E-04
0.5	2.50E-04	1.98E-04
0.75	2.58E-04	2.77E-04
1.0	2.97E-04	4.71E-04
1.25	2.97E-04	3.39E-04
1.5	3.39E-04	3.95E-04
1.75	4.14E-04	5.51E-04
2.0	5.04E-04	7.78E-04

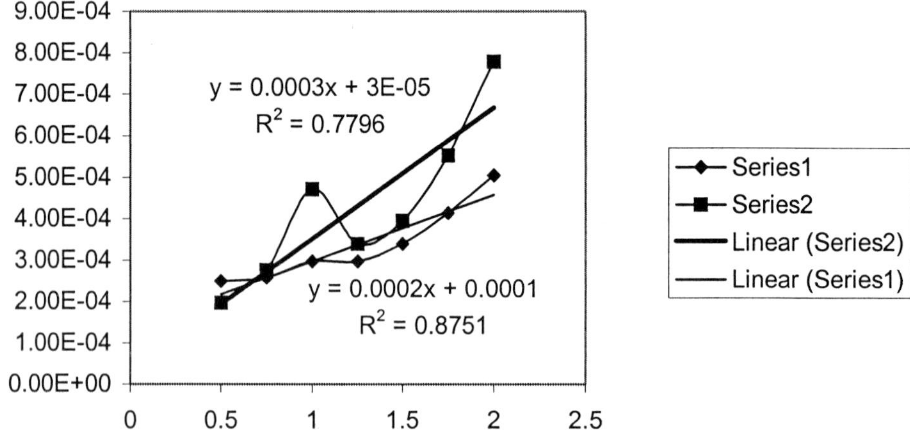

Fig. 7.3. Regression of wear data in a 4-Ball tester for oils 1 and 2. X-axis is time in hours and Y-axis is cumulative wear volume in mm^3.

be reduced compared to separate tests. Separate tests for different time intervals were conducted in the cited example due to lack of provision for periodic measurement. On the basis of available correlation the wear rates for oils 1 and 2 were 2.0×10^{-4} and 3×10^{-4} mm^3/hour respectively. A better method is to utilise the running-in equation and perform non-linear regression as reported in 5.4.1. The detailed methodology is given in the corresponding reference [5]. Such regression is possible only if there is continuous increase in wear volume with time. The procedure could thus be adapted only for oil 1. The regressed equation obtained by this procedure is

$$V = 1.83x10^{-4}(1 - e^{-4.36t}) + 1.31x10^{-4}t \qquad (7.4)$$

On this basis the results are as follows:

Running-in period 1.09 hours
Steady state wear rate $1.31x10^{-4}$ mm^3/hour
Correlation coefficient 0.962

Provided the wear scars are measured at each stage in a continuous test and adequate numbers of points are available, this method can be used to classify wear rates accurately. It may be recalled that completion of the running-in is defined as the point at which 95% of the value equals steady state wear rate.

The lower variability in the case of oil 1 suggests that this formulation is probably more effective in the running-in stage. Similar tests were conducted under different operating conditions with short duration tests with different time intervals. In all cases oil 1 showed much less random variation with time as compared to oil 2. The suggested simple linear regression does not ensure that the running-in influence is eliminated. In practical terms the effect will be substantially reduced and to that extent provides better information on wear rate. On the basis of the selected regression zone the comparison between additives will be more meaningful. Based on this approach it is possible to evolve a standard procedure for wear rate determination. Such a procedure can define the minimum number of tests to be done and the nature of regression to be done. The best approach will be to use the non-linear regression as a standard procedure. While issues of standardisation depend on the agreement between the concerned organisations, the approaches may be adapted by researchers and formulators who are interested in distinguishing between different formulations more effectively.

Large number of wear tests are reported with 4-Ball machines at different conditions. The major interest in such studies is to distinguish between different additives used in formulations. While the nature of additives is getting sophisticated the approaches to wear testing in many cases simply depend on the final wear scar diameter in a given test. The inadequacy of such an approach has been discussed above. A recent paper [6] discussed the wear behaviour on the basis of wear volume and wear coefficients. The wear volume was that obtained at the end of the test. Another paper [7] used the concept of 'delta wear' to characterise the wear behaviour. Delta wear was defined as the difference between observed scar diameter and the Hertzian diameter at the selected load. The idea is perhaps to account for the elastic recovery. On the basis of Eqn.7.2 delta wear is not proportional to wear volume and it is better to calculate the wear volume directly by this equation. An interesting paper [8] considered step load tests but characterised the wear behaviour on the basis of the step between 30 and 60 minutes only using the concept of delta wear as above. One more example of step load tests [9] is a comparison of wear scar with progressive loading up to thirty minutes only with 5 minutes steps. Influence of hexoyl borate was compared on this basis. There is an obvious need for uniform procedure to determine wear rates based on regression. While investigations are reported at few different conditions the author has not come across any detailed wear mapping for antiwear additives.

The wear tests serve two purposes. One is the specification requirement imposed for 'pass' under defined test conditions. Such tests specify a maximum allowable scar diameter. The more important utility is in screening formulations. The confidence levels in screening are important and the first step is to have an effective procedure to clearly distinguish in a given machine. It is hoped the suggested approach including wear mapping will be useful for this purpose. The issue of relating such information to practice will be considered in the later parts of the chapter.

7.3.2 Other laboratory rigs for wear evaluation

Several other test geometries can be used for evaluating wear. These include the ring-on-block test based on ASTM D 2714, as well as reciprocating rigs of different kinds. The Falex V-Block tester is also used for wear studies as per ASTM D 2670. In this machine a ratchet mechanism advances to maintain constant load on the pin. The number of ratchet teeth that advance during the testing period indicates the wear. Disk machines with a given slide/roll ratio can also be used for wear evaluation. Disk machine tests have the advantage of constant stress during operation. The problem of contact stress variations has already been considered in the first section. Most of these tests use prescribed steel specimens with given hardness and roughness. Very little information is available regarding correlations if any between different test machines and the choice is difficult. One interesting paper [10] evaluated the EP and antiwear properties of different metal based dithiophosphates as well as different types of zinc dithiophosphates with 4-Ball and FZG test rigs. It was observed that performance of different zinc salts depends not only on their chemical structure but also on the stress and nature of contact geometry in a given test machine. Such differences are expected because the gear rig operates with line contact and high Hertzian stress whereas the wear as well as EP tests in 4-Ball machines will have variable contact stress. In the antiwear tests the stresses in the contact zone as the scar develops will be lower than in FZG tester. Also the sliding speeds and temperatures involved will be different in both cases. This example shows the problem involved in selecting any particular wear test. If there is mapping of wear available at different operating conditions for various test machines an explanation is possible.

One additional factor to be taken into account in laboratory testing is the possibility of partial hydrodynamic effects that may develop in the contact. This has to be assessed by observing the changes in wear volume in conjunction with friction coefficient. A large decrease in friction with time is indicative of hydrodynamic effects. This will also be reflected in negligible change in wear volume with different time intervals and the test is no longer representative of the antiwear film alone. Theoretical evaluation of EHD influence is difficult. With the low pressures involved in the worn scar the problem will fall in the regime of rigid-variable viscosity with the added complication of ill-defined pressure distribution. The practical approach can be to select conditions that minimise such effects by reduced speeds or higher temperatures. Electrical contact resistance between the rubbing surfaces can be measured with a small applied potential in the milli volt range. A short indicates a metal contact while high resistance indicates separation by oil film. Suitable electronic circuitry can be utilised so that voltage output can be interpreted as a function of metal contact. Such approach is difficult to apply for additives as the antiwear films themselves can have high resistance. An attempt by Sethuramiah et al [11] to study possible hydrodynamic effects in a ball-on-disk machine using tricresyl phosphate as an additive has been reported. Though metal contact influence was accommodated by an empirical approach the influence of hydrodynamic effects were far less than theoretically anticipated. Thus the assumed hydrodynamic effects may as well be due to high resistance of antiwear films. The author hence considers the variation of friction coefficient is a simpler and better criterion for assessment of any hydrodynamic effects in the contact.

Most of the work being reported in the literature on wear does not take into account the surface temperature rise in contact. For example with a 4-Ball machine running at 1200 rpm under a load of 40 kg the average surface temperature rise will be 15°C for an assumed friction coefficient of 0.1 and a scar diameter of 0.5 mm. The asperity temperature rise can also be estimated for an assumed asperity contact dimension. Such an analysis will help in a better assessment of the additive.

The metallurgical aspects are also important in testing. The wear tests are done with a given steel combination. The 4-Ball machine uses AISI steel balls with Rockwell C hardness of 64-66. On the other hand in the case of the block-on-ring test the ring is made of SAE 4620 steel with a Rockwell C hardness of 58-63. The block is made of SAE 01 tool steel with a hardness of 27-33 HRC. Thus the steel combinations are different for the two machines. The wear behaviour for a given additive is expected to be different in the two cases. The question that arises is whether relative rating of additives will be same in the two machines. This question cannot be answered unless a necessary investigation is done. It is obviously preferable to have the same material combination as in the intended application. While standard tests can be conducted with standard materials for specification purposes, research investigations should simulate the required material combination whenever possible. In some machines like 4-Ball it is impractical to change materials. On the other hand in many reciprocating machines, extensively used to study ring-liner tribology, actual ring and liner pieces are adapted for testing. In many cases the purpose is in fact to evaluate material influence on wear. Also many machines are now available in which several contact geometries can be adapted. It may be mentioned that a larger number of test rigs are available for dry testing with greater degree of simulation. Blau [12] has listed the current available test rigs for dry friction conditions. The rigs are used for diverse purposes from floor friction to brake performance. On the other hand most of the studies reported on antiwear additives are confined to 4-Ball, ring-on-block and reciprocating testers. In all test machines there is a need to assess wear rate more effectively. The considerations involved are similar to those discussed for 4-Ball machine. The above discussion is limited to the issues related to better wear evaluation and the need to map performance as a function of operating conditions. The approaches still remain empirical. The practical utility of such tests has already been discussed. The nature of empirical relationships observed can, to an extent, provide an insight into the mechanisms involved as discussed in earlier chapters. The issues involved in basic modelling are elaborated in chapter 5 and hence not considered here.

The wear in real systems is complex and depends on many factors. There can be several starts and stops, operating conditions may vary over a wide range, and there can be some ingress of abrasive particles despite filtration. Another important factor is the deterioration of the base oil as well as additives with use, which again influences wear. In some formulations like engine oils the complex interactions between several additives further complicate additive action. Despite this scenario formulations with better antiwear capability will give relatively longer component life and their development is important. Two additional aspects arise with regard to development of formulations. The formulations may have to be additionally evaluated with artificial ageing to assess the real use situation. Only few investigations are available with used oils. The other issue is the response of the additive to the operating environment. From this point of view a formulation should have the least possible variation in wear rate in the required operating range. Wear determination in real systems and their comparison with laboratory rig tests is useful. But wear determination in many real systems is possible only after a long duration when measurable wear has occurred. These issues will be considered in chapter 9.

7.4 Load carrying capacity of lubricants

In the present context the load carrying capacity of a lubricant refers to the EP additive in the formulation. The purpose of EP additives and their action mechanisms have already been elaborated in earlier chapters. The present section deals with their evaluation. The test configurations used are 4-Ball, ring–on–block, and journal rotating against V- blocks. The standard test methods are listed below in Table 7.3. In addition FZG gear rig as per DIN standard is also employed to study the load carrying capacity and is adapted by ASTM and IP. As stated in the standards it is for the users to decide the issues related to correlation with practice and the acceptable levels of performance required. This clause is similar to the situation with the testing of antiwear additives.

Table 7.3
Test methods for EP properties

ASTM D 2782 Measurement of EP properties of lubricating fluids (Timken method)
ASTM D 2783 Measurement of EP properties of lubricating fluids (4-Ball method)
ASTM D 3233 Measurement of EP properties of lubricating fluids (Falex pin and Vee block methods)
ASTM D 5182 Load-carrying capacity tests for oils- FZG machine

Detailed descriptions of these methods are available in the relevant standards. The present section only discusses the main issues involved in EP testing considering the test methods as a means. The load carrying capacity, which is also called EP property, is normally the load at which a lubricant fails under prescribed conditions. The first three machines initially involve point or line contact but during test the geometry changes to that of a conformal contact. For convenience these are grouped under conformal contact machines. The gear and disk machines are classified under line contact machines.

7.4.1 Conformal contact machines

The prescribed conditions vary significantly between these test machines. Firstly in the three machines the nature of steels used are different. The sliding velocities are also different for the three machines. For 4-Ball testing the sliding speed is 0.683 m/s while it is 2.06 m/s for the Timken machine and 0.096 m/s for the V- Block machine. If it is assumed that the temperature rise in contact is related to lubricant failure directionally the failure load for the three machines should be

V- Block machine > 4-Ball machine> Timken machine

The test method prescribed for 4-Ball machine is rather complex. Firstly the test load at each stage is prescribed in such a manner that the variations are approximately logarithmic. At each stage the tests are conducted with a fresh set of steel balls and the test run for a short duration of 10 seconds. The average wear scar is measured at each stage. The testing is continued till welding is observed. Weld point is one parameter for rating. Another parameter is called the load-wear index calculated on the basis of wear scars for ten stages below welding. It is obtained by

$\frac{\sum L d_H / d_s}{10}$ where d_H is the Hertzian diameter, d_s is the scar diameter, and L is the machine load

The load-wear index is indicative of the wear behaviour under EP conditions. A higher value indicates lower wear.

Both Timken and Falex tests prescribe a run-in procedure. After the run-in the loads are increased in the prescribed fashion till scuffing is observed. Each test shall be done with fresh surfaces. The Falex test incorporates another procedure also, in which the loading is increased in steps to failure with the same surfaces. All the machines have provision for determining frictional torque although this determination is not mandatory for assessing EP properties. The machines are well designed so that loading is uniform. Material specifications as well as roughness range are rigorously followed. The Timken machine does not involve welding. The failure criterion is based on the nature and extent of scoring observed as per laid down procedure. OK load is defined as the load which is one step below the failure load. OK load may be taken as the limiting load without failure. In the Falex tester the failure is considered to occur when the pin breaks or when the load cannot be maintained.

The relative ratings will depend on the nature of the machine and operating conditions. From the available information in ASTM D 3233 the results from Falex machines correlate reasonably with 4-Ball tester.

7.4.2 Influence of operating parameters

From a fundamental point of view the EP film failure should depend on the procedure adapted. These effects are now considered with two examples. The first case is the work on a 4-Ball tester. The other is an example of scuffing tests in a reciprocating tester.

7.4.2.1 4-Ball tester

This example is based on the work of Sethuramiah et al [13]. 4-Ball tests were conducted for one minute at a speed of 0.54 m/s with a fresh set of steel balls. In each test weld load was obtained. From the measured torques with suitable springs the friction coefficient was calculated. Two types of behaviour were observed in these tests. In one case the welding occurred with a sudden transition from a relatively low friction. In another case the friction increased to a higher level followed by eventual failure with a transition from higher friction. These are termed Tr_1 and Tr_2 transitions. The behaviour depends on the nature of the additive and its concentration. It may be noted that the additives which survived Tr_1 transition showed high friction without failure for 2 to 3 stages before the weld load was reached. The stages used in these tests were 20 kg load increments. The behaviour observed is illustrated in Fig. 7.4. It was argued that the Tr_1 was related to lubricant failure temperature while Tr_2 was related to the EP film failure. The behaviour up to Tr_1 is termed the X stage while behaviour beyond the Tr_1 is termed the Y stage. The failure temperatures were calculated with the available friction coefficient and an assumed contact area based on the Hertzian contact dimension. Better estimations of temperature are possible now but the previous reported values may be used for comparison. The temperatures at Tr_1 ranged from 300 to 360°C. The temperature at Tr_2 transitions ranged from 720 to 800°C. The relatively high lubricant failure temperatures were considered to be due to the oxidised lubricant products present on the surface. The Tr_2 transitions were attributed to the EP film failure. The tests were conducted with 0.294% of sulphur concentration for all additives. Diallyl monosulphide, dibenzyl monosulphide, and diphenyl disulphide that are less reactive directly fail beyond Tr_1 indicating their inability to react fast enough to sustain EP action. At a higher concentration diphenyl disulphide became more effective showing both types of transitions. The failure loads for different additives were different and the Tr_2 failure temperatures were based on friction coefficient before failure. These temperatures were treated as 'critical failure temperatures' in the paper.

Failure temperatures alone do not offer an explanation of the mechanisms. The possible mechanisms for failure in EP conditions have been considered in section 5.5.2. The high removal rates can be due to compositional changes in the film near Tr_2 transition. In the same paper it was shown that with sulphur (0.29%) and diphenyl disulphide (1.48%) large amount of sulphur was observed on the surface just before failure at Tr_2. Oxygen was not evaluated. Oxygen availability for reaction depends on the solubility as well as its consumption due to oxidation of the fluid. These effects depend on the temperature in the contact and can change the sulphur to oxygen ratio unfavourably. Such films with changed composition can have high wear rate. The reaction rate is inadequate to cope with the situation and leads to failure. This is only a tentative explanation. Direct failure at Tr_1 on the other hand can be mainly due to the inability of less reactive additives to react fast enough to cope with the transition when the total load shifts to EP films. Yet another complication is the possibility of metallurgical changes at the high temperatures involved near Tr_2. Any softening near the surface can adversely affect the EP action. Limited study by the authors indicated that for elementary sulphur and diphenyl disulphide there is softening below the surface as evidenced by micro hardness measurements as a function of depth at one stage below welding. The reduction was nearly half the original hardness in the immediate vicinity of the surface. It may be noted that in the cited paper only softening was considered as an explanation for failure. The idea was that the harder films would not be able to adjust with the easier deformation of the substrate leading to film failure.

Some studies conducted with a step load procedure were also reported in the same paper. These studies were conducted with an initial running-in followed by one minute tests at progressively increasing loads till failure. These tests were revealing. Less active additives showed higher weld loads in comparison to standard tests. The friction coefficients were low ranging from 0.03 to 0.07 indicating partial EHD effects. The wear scars were much smoother. Thus the load carrying capacity of diphenyl disulphide at 1.48% concentration increased to 380 kg as compared to the value of 240 kg in one minute tests. Hexachloroethane with a low friction coefficient of 0.03-0.04 showed a weld load of 400 kg which is much higher than the value of 190 kg obtained in the one minute tests. The increase in weld loads in step-load tests was attributed to the hydrodynamic effects. The issue of barrier films in such cases was discussed in chapter 5.

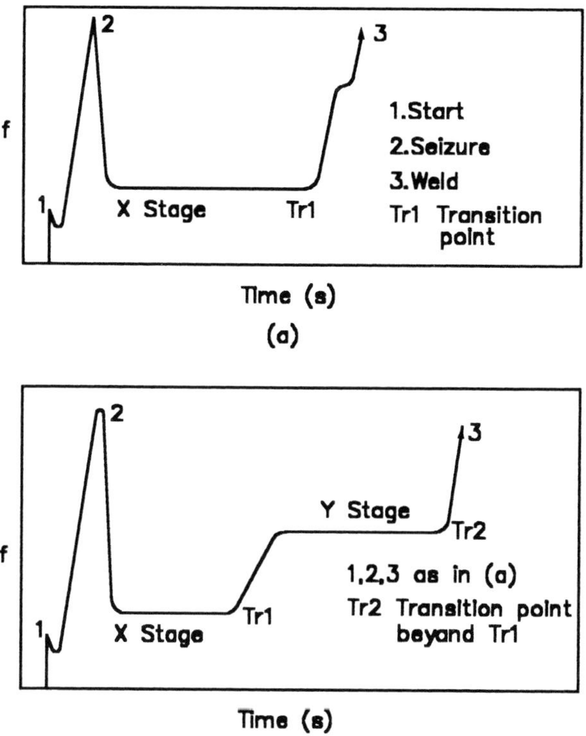

Fig. 7.4. (a)Tr_1 and (b)Tr_2 transitions observed in one minute tests in 4-Ball machine. (Reproduced from Ref. [13].

The purpose of the above example is to demonstrate the variability involved in testing based on operational conditions and to consider the failure mechanism to the extent possible. The above methods involve varying scar diameters in a test. Thus the apparent pressures based on geometric area keep varying. However as areas increase with load the pressure variations may not be large and typically for a 4-ball machine the pressures may range from 150-250 kg/mm^2. In standard tests the initial contact pressures will be high and the contact may be elasto-plastic rather than elastic. For example with a machine load of 400 kg the elastic contact diameter based on Hertz theory will be 0.642 mm. The pressure acting on the ball contact will be 473 kg/mm^2, which is significantly higher than the yield pressure. These initial conditions in tests are also expected to play a role in determining the nature of films that develop.

7.4.2.2 Reciprocating tester

It is of interest to consider the influence of rate of loading on scuffing. Gondal et al [14] conducted scuffing tests with different formulations in a SRV tester. This is a reciprocating tester, which was also used in the experiments conducted to develop wear equations discussed in chapters 3 and 5. Recently ASTM has adapted this machine to evaluate friction and wear properties of EP oils as per D 6425. A sketch of the machine is given in Fig. 7.5. The lubricants studied are given in Table 7.4. The tests were conducted at 10 Hz frequency with a stroke length of 2.0 mm. The load was increased at 14 N/s for slow rate tests while it increased at 42 N/s for rapid tests. Area contact did not produce scuffing and the tests were conducted with machined liner surfaces with chrome plated and cast iron rings. The initial contact was thus line contact. The scuffing data obtained are given in Table 7.5.

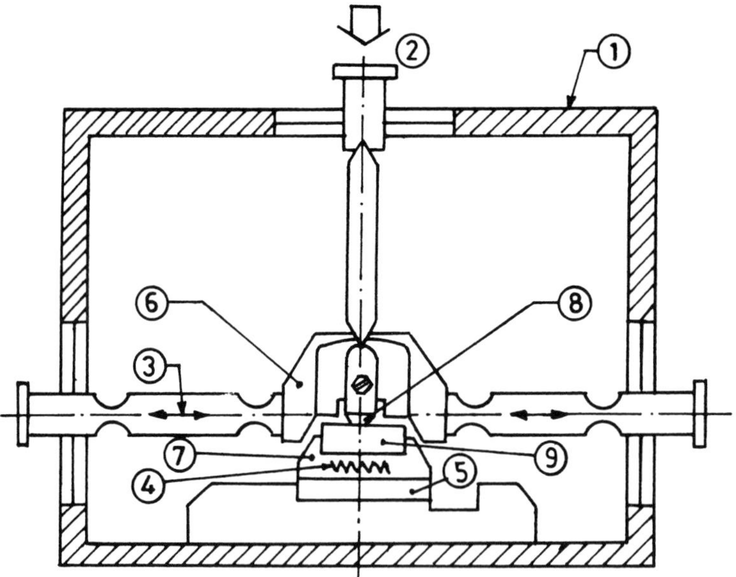

Fig. 7.5. SRV Reciprocating tester: (1) test cell; (2) loading system; (3) horizontal movement (reciprocating); (4) heating system; (5) Piezo measurement system; (6) ball holder; (7) disc specimen fixture; (8) ball specimen; (9) flat disc specimen.

Table 7.4
Lubricants studied

Oil	Description	Viscosity cSt at 40 and 100^0C		Mo, ppm	Zn, ppm
A	Base oil	64.2	7.78	0	0
A1	A + 1% MoDTP 1	-do-	-do-	86	113
A2	A + 1% MoDTP 2	-do-	-do-	380	0
B	Engine oil	129.4	13.94	0	742
B1	B + 1% MoDTP 1	-do-	-do-	86	885
B2	B + 1% MoDTP 2	-do-	-do-	380	742

Table 7.5
Scuffing tests

S. No	Contact	Type of loading	Oil	No. of tests	Scuffing load, N
1	L3	Slow	A	1	450
2	L3	Slow	A1	1	X
3	L3	Slow	A2	1	X
4	L3	Slow	B	1	X
5	L4	slow	A	1	X
6	L3	Rapid	A	5	400
7	L4	Rapid	A	2	X
8	L3	Rapid	B	1	X
9	L4	Rapid	B	1	X
10	L3	Rapid	A1	3	700
11	L4	Rapid	A1	1	X
12	L3	Rapid	A2	3	900
13	L4	Rapid	A2	1	X
14	L3	Rapid	B1	1	X
15	L3	Rapid	B2	1	X

X no scuffing
L3 cast-iron ring on flat, line contact
L4 chrome-plated ring on flat, line contact
(Tables 7.4 and 7.5 reproduced from Ref. [14] by permission of Leaf Coppin Publishing Ltd.)

The data shows that base oil alone scuffs with CI ring at 450N in a slow test while it scuffs at 400 N in a rapid test. With additive treated oils scuffing was observed only in the case of rapid test. Base oil with 1% MoDTP 1 (A1) showed failure at 700N while base oil with 1% MoDTP 2 (A2) showed scuffing at 900 N. Engine oil alone (B) and with MoDTP addition (B1, B2) did not scuff within the load range of 1200 N. With chrome plated rings no scuffing was observed with any lubricant both in slow and rapid tests. This example shows that scuffing is very much influenced by rate of loading as well as materials.

The above examples clearly bring out the fact that scuffing is influenced by the rate of loading, nature of lubricant, material combination, and test procedure. The work also points to the need for understanding film failure mechanisms by studying their detailed composition at different stages up to welding as suggested in chapter 5. If we are equipped with this information it is easier to compare various machines and to arrive at more objective test procedures. It is also important to record friction coefficient values. This will provide an insight into the partial hydrodynamic effects. Such effects may be more important when there is a run-in done before the test. Failure mechanisms have a direct relevance to the additive formulation. For example if the film failure is related to the composition changes of the film the additive chemistry has to minimise these variations. In the related area of antiwear additives effort is being made to improve performance of dithiophosphates. Based on the available understanding of the mechanisms reduced surface oxidation is desirable. Hence there are attempts to use antioxidants like copper naphthanates in the formulation as cited in 5.3.2. Such insights are possible when failure mechanisms are understood.

7.4.3 Line contact testers

Another important test used for EP evaluation is the FZG gear tester. This rig is mainly used to evaluate gear oils. This is a back-to-back gear tester in which load application is by a clutch mechanism. Each increasing load is called a 'stage' and it is common to specify load carrying capacity by stage number. The normal force on the teeth varies from 99 N to 15826 N from first to twelfth stage. The corresponding Hertzian contact pressures range from 0.146 GPa to 1.841 GPa. The test gears used are spur gears. The pinion has 16 teeth while the gear wheel has 24 teeth. The procedures to assess scuffing under the prescribed operating conditions are rather elaborate and described in the standard. Other gear rigs available include the IAE rig specified in IP and the Ryder gear rig used to evaluate gas turbine lubricants. Industry in general relies more on the FZG test for the specification of gear oils. The advantage of the gear rig is that contact stresses remain constant for a given applied load. Also it simulates at least the spur gears widely used in industry. The possible complications involved with varying stresses discussed in the previous section are avoided. The test is run at each stage for 15 minutes until failure. From a tribological point of view this means that films are progressively conditioned with each stage and will be more uniform. Failure of such films can be due to an imbalance between formation and removal rates or other reasons as discussed earlier. Attention may have to be paid to the partial EHD effects that may arise in gear rigs. If there is smoothing of surfaces due to chemical additives this factor can additionally contribute to such effects. In contrast the standard 4- Ball test involves increasing loads on fresh surfaces and running for a short duration of 10 s. The films are ill conditioned in such a case and the situation is closer to the control of fast growing metal contact area. The problem is similar with the Timken machine but since each test is run for 15 minutes there will be conditioning of the film in the later stage. The author has no clear information whether failure is basically governed by the initial severity of contact in this case. With Falex, in one procedure the situation is similar to the 4-Ball except for the test duration of one minute in each case and run-in involved. In the second Falex procedure where increasing loads are applied on the same surfaces, film conditioning occurs as in FZG rig though to a lesser degree. The author considers that these qualitative considerations are important in selecting a suitable test rig. It will be of interest to see whether the correlations between the rigs is possible by adapting different run-in and progressive loading procedures.

In the case of rear axle oils there is a need to test the oils in selected rear-axles. Such tests involve evaluation under high-speed shock loading conditions as well as low speed high torque conditions. This is because the usual EP tests cannot differentiate these properties effectively. The desired rigs are not easily available and the testing costs in the specialised laboratories are high. So an attempt was made by the author and his colleagues to distinguish the lubricants with a disk machine. The disks were run with opposite peripheral velocities to make the test conditions severe. Opposite peripheral velocities also ensure boundary conditions. Hence partial EHD effects are eliminated. Intermittent loading was applied with an eccentric mechanism. The initially conditioned run-in disks were immediately loaded to a high load. If no scuffing was observed the test was conducted at a higher load with a new pair of conditioned disks. Such testing could correlate well with the shock load test results. This was based on the testing of reference gear oils with known shock load behaviour in rear-axle rigs. The maximum load applied in the disk machine was 200 kg. It is of interest to note that if the loading was done gradually on the same disks scuffing did not occur within the capacity of the machine and distinction was not possible. This again showed the influence of film conditioning on EP action. The research carried out in this area as well as the nature of rear-axle tests with which correlation was attempted is available in publications [15,16]. It is necessary to resort to such trial and error procedures in the absence of detailed knowledge of mechanisms.

7.4.4 Practical considerations

The present approach is to specify a particular limit for EP property in a given machine. This is based on experience available with different fluids and their EP properties in a given machine. Neat drilling fluid containing EP additives for example, will have to meet a particular load carrying capacity in a 4-Ball machine. A drilling test under standard conditions may also be specified. Specification for a mild EP formulation for a reduction gear may have to typically meet minimum fourth stage pass in FZG rig. A 4-Ball test may or may not be specified for performance. Many organisations may have their own in-house rigs for evaluation though they may not be used for specification purposes. Such approaches are satisfactory within limits. Also reasonable extrapolations are valid provided the additive chemistry is essentially the same. Problems arise when new kinds of additives are to be evaluated or when there is any significant change in operating conditions or materials. It is clear that the whole approach to selection of additives is empirical with some criteria based on experience in test machines. Attempts to generalise failure criteria of mineral oils on the basis of critical temperatures was discussed in chapter 5. While there is no doubt that temperature is an important consideration in EP failure it is not a sufficient condition. A clear understanding of the mechanism of film failure is necessary as discussed earlier. The present scenario is unlikely to change if mechanistic studies are not pursued. It is desirable to conduct a detailed thermal analysis of the problem and to arrive at an agreed methodology for temperature evaluation in the test machines. The important influence of temperature on EP action can be then be clarified. The present methods assess EP action only on the basis of failure load.

7.5 Metal working lubrication

Metalworking is a vast area and this brief section is intended to look into the aspects of boundary lubrication only. Metal working operations involve major plastic deformation coupled with high temperatures. The processes may be divided into metal removal and metal forming processes. The metal removal processes can be further divided into cutting and finishing processes. Metal forming processes basically involve shaping of metal with large plastic deformation. The various metal forming operations include rolling, wire drawing, deep drawing, forging, and extrusion.

Lubricants are used to good effect in many metal removal processes. In such operations one of the important roles of the lubricant is removal of the large amount of heat generated. Oil in water emulsions are commonly used in such operations. The oil part can have boundary additives of different kinds including fat based materials. The cutting temperatures within the tool can typically range from 500 to 700°C in a lathe machine. The access of the lubricant into the cutting zone is difficult and is probably through capillary action. Also how an organic material can function under the high temperatures is not clear. In practice it is known that boundary additives are effective and used extensively. It is easy to simulate cutting, grinding and finishing operations for evaluating different formulations. In severe operations that involve high temperatures, and limited access, as in drilling of hard steels neat cutting oils with EP additives are used. Simulated tests for such cases involve measurement of drill torque and the number of holes that can be effectively drilled. An interesting example of such studies with vegetable based cutting oil may be cited [17]. Several other issues like corrosion of the work piece, toxicity, and related issues have to be additionally taken into account. 4-Ball EP performance level may also be specified in some cases.

7.5.1 Metal forming operations

Many metal forming operations involve mixed lubrication conditions with partial hydrodynamic effects. The simplest test that can be used to study a lubricant is a ring compression test. Compression of a lubricant coated surface under plastic deformation changes the ring dimensions depending on the interfacial friction. A complex test can involve a laboratory rolling mill to simulate high speed cold rolling. Such a rig can be very expensive. Several levels of simulation are possible between the two extremes considered. Rigs being used include small scale low speed rolling, wire drawing, cup drawing, forging, and extrusion. The problem with the smaller rigs in many cases is that they cannot simulate the large-scale industrial operations and the available theoretical base is not adequate to extrapolate the results to real conditions. An example may be cited from the important area of cold rolling. While film thickness for cold rolling conditions can be predicted for smooth surfaces the modelling of friction presents major problems. A recent attempt to predict friction based on empirical modelling [18] is indicative of the complexities involved. The problem of friction is more complex in deep drawing where different zones will have different friction coefficients. The importance of interfacial friction arises in the FEM analysis of the deformation processes necessary for design. The boundary conditions to be defined necessarily involve interfacial friction.

The effect of boundary lubrication on roughness and surface integrity of the work is not governed by friction alone. The nature of surface interaction with the boundary lubricant is obviously important. As many tests normally used involve mixed lubrication it is difficult to isolate boundary effects. A fundamental approach should involve the capability to investigate boundary effects under plastic deformation with a wide range of sliding speeds. Such mapping will considerably aid in understanding boundary lubricant effects that can be integrated into the overall assessment of the lubricant. A new approach to study the boundary lubrication problem in plastic deformation developed by Banerjee et al [19] is described below. The discussion here is confined to the main concepts involved.

7.5.1.1 Evaluation of boundary effects

Investigation of boundary effects in plastic deformation at a laboratory level needs a rig that can operate over a wide range of loads and sliding speeds. A simple and effective idea used for this purpose was an oblique plastic impact. A steel ball impacts an oblique plane of the work piece B. The ball indents the work piece with plastic deformation and also slides along the plane. The initial velocity component along the plane reduces due to friction as the ball slides. Finally the ball rebounds with the final velocity components and makes a second impact on another plate H. The onset of impact and rebound are illustrated in Fig. 7.6. The rig is illustrated in Fig. 7.7. By modelling the flight from rebound to second impact as a projectile motion and measuring the time and distance the velocity along the plane at the point of rebound can be calculated. It can be shown that the average friction coefficient at the interface during sliding is

$$f = (V_{H1} - V_{H2}) / (V_1 + V_2) \tag{7.5}$$

where

V_1 and V_2 = Velocity components perpendicular to the plane at onset and rebound

V_{H1} and V_{H2} = Velocity components along the plane at onset and rebound

The velocity components are also shown in Fig. 7.6. The detailed theoretical consideration is available in the reference. From the available geometry it is obvious that a wide range of impact loads and sliding velocities are possible by simply altering the impact height and oblique angle.

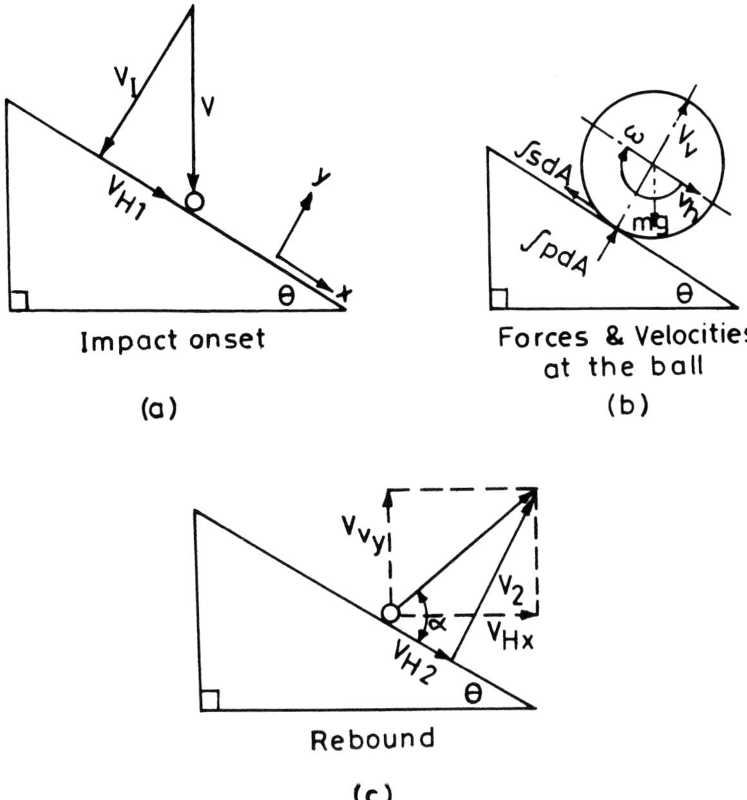

Fig. 7.6. Impact and rebound components of the ball. (a) Velocity components at onset. (b) Forces and velocities for the ball. (c) Rebound velocity components. (Reproduced from Ref. 19 by permission of STLE)

The lubricant is applied to the surface of the low carbon steel piece. The nominal thickness of this layer was 1.0 μm to minimise possible hydrodynamic effects during sliding. It was assumed that the contact essentially operated in boundary lubrication regime. This assumption is supported by the work of Imado et al [20] who have observed significant metal contact when traction fluids were studied under plastic impact with sliding. On the other hand when the oblique impact involved elastic contact there is evidence of full separation as reported by Jacobson

[21]. The steel surface had R_a roughness of 0.25 μm. Different lubricants were compared with this technique. Two oblique plane angles of 30 and 45° were used. The fall heights were varied between 0.85 to 2.20 m resulting in an overall variation of 2.04 to 4.64 m/s in the initial sliding velocities. An EN 31 steel ball with a diameter of 0.167 m was used for the impact tests discussed below.

The lubricant efficacy was evaluated on the basis of morphology of the crater surface. The surfaces were mainly observed with an optical microscope. Some craters were also examined by SEM. Typical difference between effective and ineffective lubricants is the extent of surface flow caused by asperity interaction.

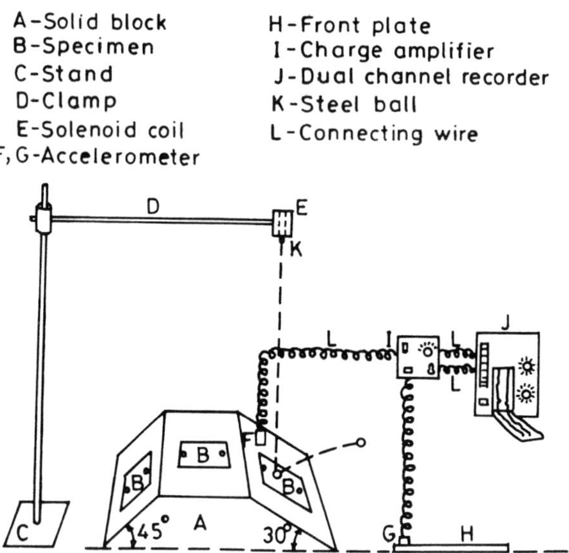

Fig. 7.7. Test rig assembly. (Reproduced from Ref. [19] by permission of STLE)

Fig. 7.8a shows the observed crater surfaces for dry, n-hexadecane, stearic acid, and oleic acid at x 50 magnification. The surfaces were observed at the centre of the crater except for n-hexadecane. With n-hexadecane the surface observed was slightly away from the centre and shows the border between impacted and original suface. The test conditions involved 1.5 m fall height and an oblique angle of 30°. The fatty acids were used at a concentration of 5% in hexadecane. In the case of dry impact as well as hexadecane alone, there is significant plastic flow as well as evidence of material pull out leaving shallow pits. The flow is such that the original grooves in the surface are covered in many zones. Similar observation may be

Fig. 7.8a Optical photomicrographs of impacted craters: (i) dry contact (ii) n-hexadecane (iii) stearic acid and (iv) oleic acid. (Reproduced from Ref.[19] by permission of STLE)

made with hexadecane alone though there was some reduction in the scale of deformation. With oleic acid the clear pattern of original striations can be seen. In this case the plastic deformation does not extend to the valleys and also there is no evidence of metal pull out. This is interpreted as due to effective boundary lubrication that prevents asperity level adhesion and prevents the large scale flow observed with hexadecane and in dry sliding. The behaviour of stearic acid is not as effective as oleic acid. In some zones flow similar to dry case can be seen. Such zones may arise due to partial failure of the film at some contacts. In Fig. 7.8b two SEM micrographs are included for typical surfaces. These micrographs at high magnification are given for dry and oleic acid impact. The differences in surfaces are more obvious in these photomicrographs. Palm oil used neat also showed very effective boundary action and the observed surface is not included in the figure. It is reasonable to consider that the large variations in crater morphology with different lubricants is due to boundary effects. This additionally supports the assumption that boundary lubrication was involved in the contact. More detailed consideration with additional lubricants is available in the cited reference [19].

The friction coefficient has been found useful mainly in deciding the effective operation zone for a lubricant based on transition. The friction coefficient was related to $P(V_{Have})^{0.5}$ where P is the normal load and V_{Have} is the average of the velocities along the sliding plane at onset and rebound conditions. The transition point in friction determines the operating parameters below which the lubricant is effective.

The above study clearly brings out the possibility of evaluation of boundary effects by the simple plastic impact technique. Various studies at different temperatures and wider operating conditions are possible. Some work was also conducted with regard to the influence of polar compounds on surface hardness. It is of interest to observe that at identical impact conditions the crater depths were different for different lubricants. Surface hardness was inversely related to crater depth. There is thus directional evidence for the influence of polar molecules on strength properties. Investigators in the past have reported such influences on single crystals [22]. Plastic deformation with some level of orientation in the direction of sliding may promote such effects. It is also possible to study the surface reactions by FTIR or other techniques. In the present case metal transfer to the ball was not investigated. Such studies can be useful to assess the lubricant failure. Investigations done with regard to lubricant-metal interactions under plastic deformation at a fundamental level are limited and the author considers that there is a need to strengthen this area.

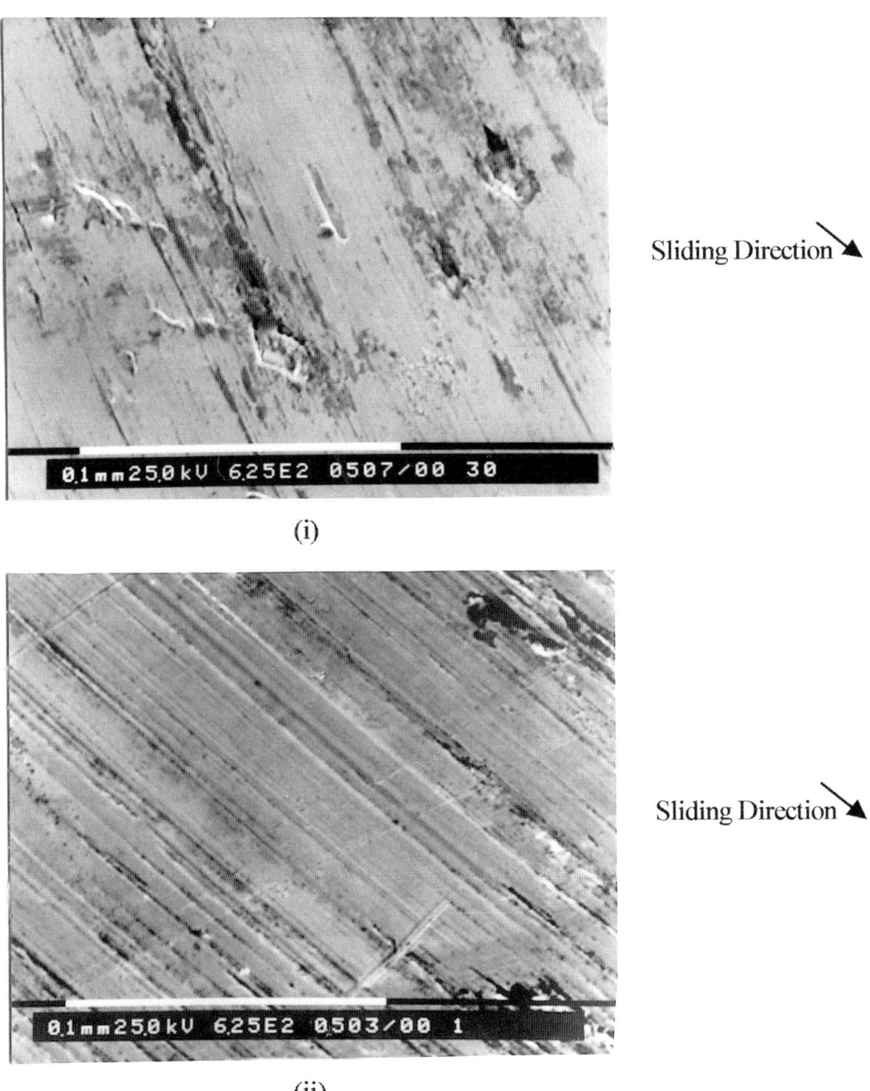

Fig. 7.8b. SEM photomicrographs of impacted craters: (i) dry contact and (ii) oleic acid. (Reproduced from Ref. [19] by permission of STLE)

7.5.2 Other aspects

The above consideration of metalworking lubricants is limited and only an approach to study boundary aspects was given. The lubricants in practice are varied and complex. The available literature mentions the lubricants in general terms and it is difficult to give a detailed account of the products involved. The available information is given in [23]. The complexity can be appreciated with one example of the cold rolling of steel. In cold rolling of steel meta-stable emulsions are common. They should separate fast enough in the roll bite to provide essentially an oil layer. This 'plate out' characteristic has to be optimised for different operations and this itself involves difficult technology. Besides boundary additives several other additives in smaller quantities like brighteners, corrosion inhibitors are also added. Even in cold rolling the interfacial temperatures can range from 500 to 600°C. The role of lubricant is probably related to its survival in the short contact time. The decomposed lubricant on the surface should be easily removable without staining. The roughness and brightness of the finished products is important and is related to the initial roughness of the roll and work as well as operating conditions in a complex manner. In hot rolling of steel also lubricants are used and include all sorts of materials starting from fats to solid lubricants. Several other ideas used in metal forming include coating of the sheet with soft metals as well as 'conversion coatings' that are obtained by phosphating. Such coatings are used in conjunction with solid or liquid lubricants. Existing practices and formulations will have their limits of applicability. For example in the high speed rolling of steel beyond 12 m/s the existing lubricant technologies were inadequate and new approaches became necessary. Synthetics and semi-synthetics with recirculation is one of the effective solutions available today. Ability to screen boundary performance in such cases becomes important and the approach suggested above can be useful in screening at high sliding velocities.

The oblique plastic impact technique involves very short contact duration of the order of 10^{-6} seconds. At this stage it is not known whether the technique can be applied to other evaluations. With more severe operating conditions the technique may be adaptable to evaluate the shock loading capability of rear-axle lubricants.

References

1. C. N. Rowe, Lubricated wear, in M. B. Peterson and W. O. Winer (eds.), Wear Control Handbook, ASME, New York, 1980, 143-160.
2. M. B. Peterson, Design considerations for effective wear control, in M. B. Peterson and W. O. Winer (eds.), Wear Control Handbook, ASME, 1980, 451-456.

3. G. Salomon, Failure criteria in thin film lubrication-the IRG programme, Wear, 36 (1976) 1.
4. P. Shrivastava, Wear characteristics of running-in oils, M. Tech thesis, IIT Delhi, 1997.
5. R. Kumar, B. Prakash and A. Sethuramiah, A systematic methodology to characterise running-in and steady state wear processes, Wear, 252 (2002) 445.
6. R. W. Hein, Evaluation of bismuth naphthanate as an EP additive, Lub. Eng., 56 (11), (2000) 45.
7. D. E. Weller and J. M. Perez, A study of the effect of chemical structure on friction and wear: Part 1- Synthetic ester based fluids, Lub. Eng., 56 (11), (2000) 39.
8. J. M. Perez, D. E. Weller and J. L. Duda, Sequential Four-Ball study of some lubricating oils, Lub. Eng., 55 (9), (1999) 28.
9. Z. S. Hu, J. X. Dong and G. X. Chen, Synthesis and tribology of aluminium hexoxylborate as an antiwear additive in lubricating oil, Lub. Sci., 12 (1999) 79.
10. M. Born, J. C. Hipeaux, P. Marchand and G. Parc, The relationship between chemical structure and effectiveness of some metallic dialkyl and diaryl dithiophosphates in different lubricated mechanisms, Lub. Sci., 4 (1992) 93.
11. A. Sethuramiah, V. P. Chawla and C. Prakash, A new approach to the study of the antiwear behaviour of additives utilising metal contact circuit, Wear, 86 (1983) 219.
12. P. J. Blau, Friction and Wear Transitions of Materials, Noyes Publications, New Jersey, 1989, 33.
13. A. Sethuramiah, H. Okabe and T. Sakurai, Critical temperatures in EP lubrication, Wear, 26 (1973) 187.
14. A. K. Gondal, B. Prakash and A. Sethuramiah, Studies on the tribological behaviour of two oil-soluble molybdenum compounds under reciprocating sliding conditions, Lub. Sci., 5 (1993) 337.
15. V. K. Jain, V. P. Sharma and A. Sethuramiah, Application of disc machines for testing the shock loading capability of hypoid and industrial gear oil, Trib. Int., 19 (2), (1986).
16. A. Sethuramiah and V. K. Jain, Evaluation of rear axle extreme pressure (EP) lubricants by a disc machine, Wear, 52 (1979) 49.
17. W. Belluco and L. De Chiffre, Testing of vegetable-based cutting fluids by hole making operations, Lub. Eng., 57(1), (2001) 12.
18. H. R. Le and M. P. F. Sutcliffe, A semi-empirical friction model for cold metal rolling, Trib. Trans., STLE, 44 (2001) 284.
19. R. K. Banerjee, C. R. Jagga and A. Sethuramiah, Friction and surface morphological studies in plastic deformation under boundary lubrication condition, Trib. Trans., STLE, 44 (2001) 233.
20. K. Imado, H. Miyagawa, A. Miura, N. Ueyama, and H. Fujio, Behavior of traction oils under impact loads, Trib. Trans., STLE, 37, (1994) 378.
21. B. O. Jacobson, Rheology and Elastohydrodynamic Lubrication, Tribolgy series, 19, Elsevier, 1991, chapter 15.

22. M. Ciftan and E. Saibel, Rebinder effect and wear, Wear, 56 (1979) 69.
23. J. P. Byers (ed.), Metal Working Fluids, Marcel Dekker, New York, 1994.

Nomenclature

d	ball scar diameter
d_H	Hertzian diameter
f	coefficient of friction
H	hardness
K	wear coefficient
L	machine load in 4-Ball machine
P	normal load
t	test duration
Tr_1, Tr_2	temperatures at which transition in friction takes place
V	wear volume
V_{H1}, V_{H2}	velocity components along the oblique plane at onset and rebound
V_{Have}	average of the velocities along the plane, $(V_{H1} + V_{H2})/2$
V_1, V_2	velocity components normal to the plane at onset and rebound

8. Fatigue and wear in mixed lubrication

8.1 Introduction

Mixed lubrication is receiving increasing attention which is focussed mainly on EHD contacts. The theory of mixed lubrication is considered in the first section. The author has attempted to explain the main concepts based on physical reasoning. While theoretical developments are advanced, experimental verification of the ideas is limited. Experimental aspects are considered in the second section. The practical utilisation of the modern theory is illustrated with the example of artificial joints and included as part of the second section.

Fatigue leading to pitting is an important consideration in many EHD contacts. Pitting in rolling/sliding contacts is considered in the third section. A brief coverage is first made of the interacting parameters involved in pitting. This is followed by a practical consideration of pitting in rolling element bearings and gears.

The lubricant-metal interaction effects on pitting are considered in the fourth section. This is a research area and the available understanding based on laboratory evaluations is presented in the first part. The next part is devoted to some pertinent observations on the mechanisms. These observations bring out the importance of competition between crack growth and wear.

The fifth section deals with the problem of wear in rolling/sliding contacts. This section is treated as an extension of boundary lubrication dealt with in earlier chapters. The contact models are first considered followed by an example of experimental work.

Contact fatigue and mixed lubrication are specialised areas with voluminous literature. The author has presented a concise treatment of the main ideas. It is hoped that such coverage will provide sufficient insight to appreciate the role of lubricants in pitting and wear. Detailed theoretical approaches are available in the cited references.

8.2 Theory of mixed lubrication

Mixed lubrication may be defined as a situation in which the load is shared by the asperities and the fluid film. The conventional approach is considered first. This is followed by a discussion of the recent developments in mixed lubrication.

8.2.1 Conventional approach

Mixed lubrication regime may be defined in terms of the film thickness to the composite ratio called Λ ratio as considered in chapter 1. The other way of visualising it is in terms of friction coefficient. As considered in chapter 1 the film thickness in a journal bearing is a function of the Sommerfeld number, $(\eta N/p)(r/c)^2$. For a given bearing with a fixed r and c it follows film thickness will be related to $(\eta N/p)$. This term for convenience is now termed S. If the friction coefficient f is now plotted as a function of S the relationship obtained is schematically shown in Fig. 8.1.

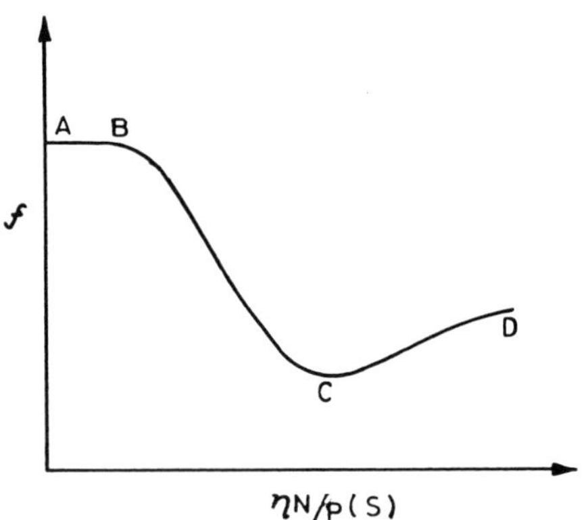

Fig. 8.1. Schematic relation between S and f for a hydrodynamic bearing.

This is normally referred to as the Stribeck curve. In this curve the behaviour to the right of the minimum follows the hydrodynamic regime with no asperity contact. The film thickness increases with S and the viscous friction increases. To the left of the minimum the film thickness is inadequate for complete separation and the load is increasingly shared by asperities as S decreases. Finally at B the total load is supported by the asperity contacts and f remains nearly constant from B to A. The region C to B is designated as the mixed lubrication regime. Typical f values in the boundary regime are 0.08 to 0.1, while typical values in hydrodynamic regime are less than 0.01.

In usual EHD contacts operating in Elastic-Viscous regime the film thickness varies with $(\eta\, u)^{0.7}$ as can be seen from Eq.1.28. The load dependence is small and may be ignored. Several modified Stribeck curves have been drawn for EHD contacts to visualise the regimes. The traction varies in a complex manner in EHD contacts due to the non-Newtonian behaviour of the lubricant. Hence the variations in friction coefficient will be specific to the lubricant used. The author considers it better to define the various regimes on the basis of the Λ ratio. The concept is common to both hydrodynamic and EHD contacts Recalling from chapter 1 the regimes based on Λ ratio are:

Hydrodynamic/EHD	> 3.0
Mixed	0.5 to 3.0
Boundary	< 0.5

Mixed lubrication is more important in EHD contacts as film thickness in such contacts is usually small. In steady running the desirable goal is to have minimum possible asperity interaction.

8.2.2 Recent developments

The above approach to mixed lubrication based on the Λ ratio needs the value of film thickness. In the simple approach the film thickness is calculated on the basis of smooth surfaces. When film thickness is of the same order as roughness, the roughness will surely affect the flow of the fluid in the contact. Early efforts in this direction involved the solution of Reynolds' equation modified for the roughness effect [1,2,3]. The average flow model due to Patir and Cheng [3] was able to quantify the roughness effects on the basis of orientation. Their model predicted highest film thickness for transverse roughness as compared to isotropic and longitudinal roughness. Transverse roughness refers to the case in which the roughness is predominantly oriented in the direction perpendicular to the entraining direction. Real roughness is made up of several peaks and valleys. In reference [3] the degree of orientation was quantified in terms of the ratio of autocorrelation functions in the two directions. While asperity contacts can occur in this model their influence was considered only in terms of change in flow factors. The influence of deformation on roughness was not considered.

From the important beginnings cited above, the study of the roughness effect has advanced conceptually and mathematically. Firstly there was a realisation of the fact that as asperities approach each other there will be local pressure fluctuations as well as deformation. These effects in turn generate separating film at the asperities. This domain of lubrication is now referred to as micro EHD (MEHD) lubrication. MEHD may be viewed as a positive response of the asperities to prevent direct contact. The fluid flow at the asperity level helps the overall macro flow in the contact and global fluctuations in film thickness are reduced. The pressure fluctuations at the asperity are, however, large and they may have a negative effect in terms of fatigue. The solution to such problem involves advanced fluid dynamics. Several papers with increasing levels of sophistication are now being published in this area some of which may be cited [4,5,6,7]. Most of the solutions so far attempted consider sinusoidal roughness that moves in the contact zone or is stationary with reference to the moving rigid smooth surface. Recently some models considered three-dimensional roughness [8]. Spikes [9] presented a useful overview of mixed lubrication where theoretical and experimental issues are discussed.

The influence of asperity deformation was theoretically evaluated by Kweh et al [4] and is shown here as an example. The authors investigated the influence of a sinusoidal roughness in a heavily loaded contact and illustrated in Fig. 8.2. The maximum contact pressure between the disks was 1.06 GPa with a mean sliding velocity of 24.87 m/s. One smooth surface rotated against the transverse roughness. The point of interest is that film thickness calculated at different operating temperatures, curves a, b, and c, are nearly flat. This means the sinusoidal roughness shown along the X-axis is completely flattened. The amplitude of the roughness in this case was 0.25 μm. The other aspect that can be seen is the significant pressure fluctuations due to asperities as compared to the smooth Hertzian pressure variation marked d. The values obtained theoretically depend on the various assumptions made, but the present example serves the purpose of visualising the strong influence of MEHD lubrication.

The above studies were mainly confined to the situations where there are no asperity contacts. The film thickness in MEHD does not depend only on the pressure fluctuation and deformation. It also depends on the shear behaviour of the film. Theoretical modelling of the shear behaviour depends on the rheological model selected as well as the temperature involved. The film thickness involved in MEHD at the asperities is at the nano level when conditions become severe. This thickness is similar to the dimensions of typical boundary films. Thus it is difficult to separate the two effects. This also has a bearing on rheological models. The available models are based on bulk fluid flow and their applicability to very thin films can be questioned. Such problems are very difficult to resolve. These issues will also have a bearing on the scuffing models being developed and discussed in chapter 5.

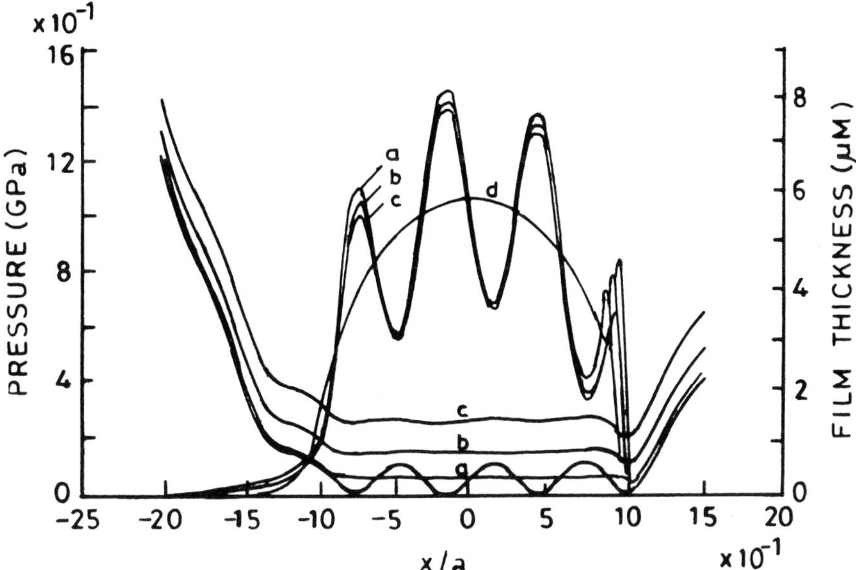

Fig. 8.2. Longitudinal centreline pressure distributions and film thickness profiles for d = 0.25 μm. (a) 140°C; (b) 80°C; (c) 50°C; (d) d = 0, 140°C. (Reproduced from Ref. [4] by permission of ASME)

The asperities eventually make contacts due to their inability to sustain full MEHD effect. Some asperities will sustain the MEHD effect while some others will make contact. Such problems are again being modelled by considering load sharing between asperities and fluid films [8]. There are theoretical attempts to re-define Λ ratio on this basis.

Another issue of relevance is the composite roughness. When two rough surfaces are in relative motion their contact is simulated by an equivalent surface moving against a rigid flat surface as considered in chapter 1. Such an approach is commonly used in modelling surface contact. In some cases there can be asperity level conformity between the surfaces with peaks of one surface matching with the valleys of the other surface. In such a case the composite roughness cannot be defined on this basis. This aspect will be considered in the next section with an example.

8.3 Experimental considerations

8.3.1 Film thickness

The significant advances in theory need to be backed by experimental verification. First consider MEHD without asperity contact. Any verification of the theory demands an ability to measure nano level film thickness. Film thickness measurements are possible by electrical and optical techniques. The commonly used electrical techniques are based on capacitance and eddy current. Such techniques will have a resolution of the order of a micrometer and are unsuitable for measuring very thin films. Capacitance techniques were used in the past to measure macro EHD film thickness and served the useful purpose of verifying the theoretical predictions. Eddy current techniques are more common for measuring thicker hydrodynamic films. An effective optical technique by light interference was first developed by Gohar and Cameron [10] and is now widely used by several researchers to measure EHD film thickness. This technique can be applied only when one of the contacting surfaces is transparent. A glass or sapphire disk is used with a semi-reflective chromium coating facing the rotating steel ball. When light is shown through such a contact there is reflection from the coated surface and the steel ball. The reflected beams are out of phase and the resulting interference fringes form the basis for film thickness measurement. The minimum thickness that can be measured by such a technique is around 100 nm. A modified technique with a silica spacer that can measure film thickness as low as 5 nm has been developed and reported by Johnston et al [11]. The composite roughness involved in the contact was 11 nm and the authors assumed that this roughness may be effectively flattened due to micro EHD effects and film thickness measurements of 5 nm are meaningful. With this technique influence of additives on the film thickness is also being studied at the nano level [12] and it is hoped that the influence of boundary and MEHD films may eventually be clarified.

The experimental work of Kaneta et al [13] exemplifies the roughness influence with single bump as well as transverse and longitudinal patterns. They used the conventional interferometry techniques to measure film thickness. These studies directionally support the existing theoretical ideas. For example, their results for a transverse asperity showed minimum deformation in pure rolling as compared to rolling/sliding contact. This is in accord with the present theoretical understanding. The minimum thickness that can be measured with such a technique will be about 100 nm and so verification cannot be extended to zones involving nano level thickness. The influence of real surface roughness is yet to be clarified experimentally.

8.3.2 Asperity contact

Asperity contact through the films is of practical importance. Precise identification of MEHD film failure is an experimentally difficult task. The problem of asperity contact has been studied over a long period of time through observations of friction and metal contact as a function of operating conditions. A better insight into the problem can be achieved when friction and metal contact are simultaneously observed. As it is not easy at present to estimate micro effects for real surfaces the observations are generally based on Λ ratio calculated on the basis of smooth surface film thickness. The metal contact circuit was first reported by Furey [14] and later used with and without modifications by several others. The main concept involved is illustrated in the circuit given in Fig. 8.3.

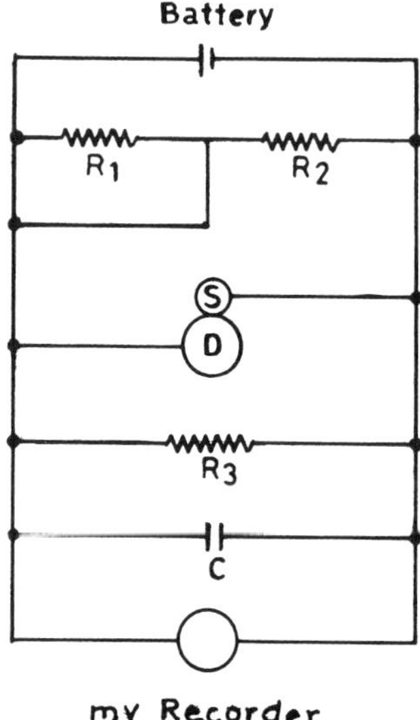

Fig. 8.3. Metal contact circuit

A small DC potential of 10-20 mV is applied across the contacting surfaces by suitably selecting the R_1/R_2 ratio. In the lubricated contact the lower disk D rotates against the stationary ball. The disk is insulated from the shaft while the ball is insulated from the holder. The leads from the rotating and stationary contact are connected to a milli volt recorder via RC circuit. When an oil film separates the sliding contact, the full applied potential will be the output. On the other hand when there is asperity contact there is shorting and the potential is nearly zero. The RC circuit modulates these variations and an average voltage is recorded with time. The actual contact situation can be seen by connecting a CRO across the contact. In such a case the fast changes from zero to full potential can be seen. The main problem of the circuit is that shorting can occur with one or more contacts. Thus the circuit does not provide information on the actual number of contacts. But statistically the probability of contact increases when more asperities tend to contact. The average output increases with decreasing contact. The nature of the circuit used is such that the variation is non-linear. The usual approach is to relate the milli volt output to the contact percentage or separation percentage. The overall response of the circuit depends on the resistance and capacitance values selected. A modified circuit [15,16] can be used to obtain linear response instead of a non-linear response. In the modified circuit square pulses are generated at high frequency. A comparator allows the pulses to pass through a gate only when a metal contact is detected. The pulses are then integrated over a selected short time interval and displayed as output voltage. The metal contact circuit is now usually referred to as the electrical contact resistance (ECR) technique.

One recent example of the use of ECR may be cited. Lugt et al [17] studied the influence of running-in in mixed lubrication in a disk machine. They studied the influence of running-in under pure rolling and with a small sliding component. The completion of running-in was based on the selected threshold value for electrical contact resistance. The running-in was first completed in pure rolling. When the operation was then shifted to rolling/sliding condition the surfaces had to be run-in again to achieve completion. To ensure completion of running-in for both conditions, the process had to be repeated by shifting from one condition to the other. This work showed that completion of running-in achieved at one condition might not be adequate for another condition.

The detection of metal contact depends on the low resistance assumed at the contacting asperities in comparison to the very high resistance of the oil film. This assumption is normally valid. But cases can arise where coherent oxides and/or reaction films are involved at the surface. In such cases there can be high resistance due to the films that may be interpreted as the oil film by the circuit. Also when nano level films are involved they may conduct electricity due to the tunnelling effect. In such cases though a separation exists between the two asperities it will be interpreted as a contact. Such considerations may become important if ECR is to applied for detecting MEHD film failure. Any fine-tuning of the circuit will depend on the resistance of the now conductive nano films.

8.3.3 Asperity level conformity

As introduced in the sub-section 8.2.2 asperity level conformity refers to a situation in which there is matching at the asperity level. Such conformity may extend throughout the surface or may be localised and has been reported in the literature [18,19]. A detailed investigation of this aspect was conducted by Tyagi and Sethuramiah [20,21]. The tests were conducted in a disk machine with a cast iron and aluminium alloy disk pair with two different lubricants. The disks rotated in opposite directions at 2000 and 2250 rpm resulting in a relative sliding speed of 10 m/s. The upper aluminium disk rotated at the lower speed of 2000 rpm. One example of conformity obtained with diester is shown in Fig. 8.4. The surfaces obtained involved four step load tests of 5 minutes each followed by prolonged running for 10 minutes at 294.30 N. The surfaces clearly show the peaks of one surface matching with the valleys of the other. The roughness was measured leaving one mm from the edge. The measurements were done in the transverse direction. To represent contact condition the upper disk profile was inverted. The matching occurred throughout the contact and similar patterns were observed all along the surface.

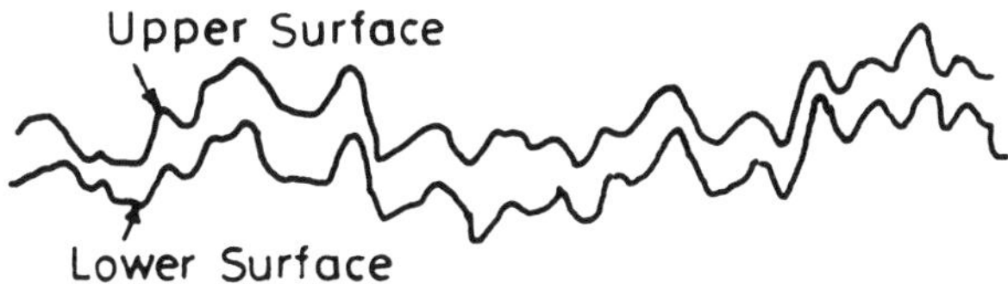

Fig. 8.4. A case of conforming surfaces. (Reproduced from Ref. [20])

The influence of roughness in such a contact cannot be treated in terms of the usual composite roughness σ^*. It may be recalled from chapter 1 that $\sigma^* = \sqrt{\sigma_1^2 + \sigma_2^2}$ where σ_1 and σ_2 refer to the roughness of the two surfaces.

The probability of contact for a given film thickness will now be far lower than that based on σ^*. The effective roughness will be composed of the difference in heights at each point. This roughness profile, called composite profile was generated on the basis of digitised values of the filtered roughness. An example of the individual profiles and the composite profile are given in Fig. 8.5 for the diester lubricant called oil B. The rms value of the composite profile, rms_z, was 0.22 μm as compared to that based on random contact, $rms_{1,2}$, of 1.35 μm. The value of 1.35

μm is what was obtained on the basis of the equation above. The cross-correlation value of 0.96 mentioned in the figure needs an explanation. The roughness profiles were measured with nominal relocation. This may not exactly correspond to the contact situation. Cross-correlation was done by slight shifting of one profile

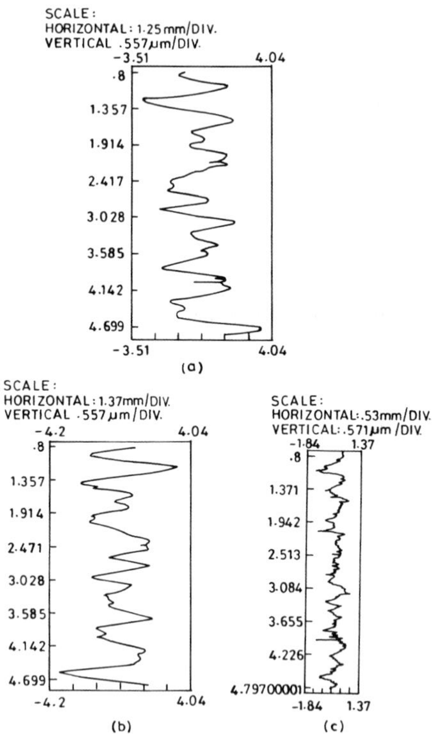

Fig. 8.5. (a) Filtered R-profile, CI disc step-load test, 294.30 N, oil B; (b) Filtered R-profile, Al disc step-load test, 294.30 N, oil B; (c) Composite profile of (a) and (b), cross-correlation coefficient = 0.96, r.m.s.$_{1,2}$ = 1.35 μm, r.m.s.$_z$ = 0.22 μm.
(Reproduced from Ref. [20])

relative to the other in the computer. The relative location of profiles at which the cross-correlation function reached a maximum value was obtained. This position was taken to be the actual contact between the surfaces. The maximum value of the function was 0.96 in the given example.

The importance of conformity can now be appreciated. For example with an assumed film thickness of 1 μm the normally calculated Λ value will be 0.74 suggesting significant metal contact. But the Λ value based on composite profile will be as high as 4.55 suggesting negligible contact. A limited attempt was made to study the onset of metal contact based on the Λ value for the composite profile. The values for onset of contact obtained for two cases were 2.11 and 2.26. Normally the onset of metal contact is expected for Λ value around 3.0. The observed values were lower, and detailed investigation was not conducted in this regard. It may be observed that the film thickness calculation was based on smooth surfaces. It is not known whether there was any increase in film thickness due to the transverse roughness generated. The details including observations on friction coefficients are given in the cited references.

The mechanism of development of such conformity is not clear and may involve micro abrasion by third bodies generated in the contact. But if it is achieved it is expected to reduce metal contact significantly and the influence can be more important than MEHD effects. Promotion of such effects by chemical additives or other means can be practically useful. Also it may be worthwhile to re-examine the concept of equivalent composite roughness, even if there is no matching. Surfaces can have random or oriented roughness and it is necessary to define the conditions in which the normally used concept of equivalent roughness against a smooth rigid body is valid.

8.3.4 Advances in practice

Improved mixed lubrication is a desirable goal to be achieved in practice. The general aspects are well known and used extensively to improve practice. The modern theoretical developments are not yet widely used in practice. Some of the theoretical concepts have been utilised in the EHD analysis of artificial joints. One interesting example is the ceramic-polymer joint that is briefly considered here. For the present purpose the artificial joint is simply considered a ball and socket joint from the lubrication point of view. Relevant references cited in chapter 6 provide detailed information about these joints. The problem is to have minimal contact between asperities when lubricated by the synovial fluid. There are two basic types of material combinations being used for such joints. One type of joint involves a metal or ceramic ball sliding against UHMWPE. The polymer is attached to a metallic backing. Polymer being an elastic material, significant elastic deformation at the asperity level is expected. The solution of such a problem needs incorporation of the time dependent variation of asperity heights into the Reynolds' equation. The theoretical evaluation [22] showed that considerable MEHD support is available in such contacts when calculations were based on transverse sinusoidal roughness. This translates into increased life of the joint as asperity contact is reduced. This is an interesting example of the use of MEHD lubrication in practice.

Another approach to reduce asperity contact is by making the roughness very low. This approach is now being adopted for metal-metal, metal-ceramic and ceramic-ceramic joints of more recent origin. The film thickness calculation is direct by using the applicable Elastic-Isoviscous equation [23]. The ball and socket joint operates with small clearances. For such conditions the elastic effects are high enough but the influence of the pressure on viscosity is negligible. Hence the pressure effect on viscosity is neglected. The calculations at the applicable conditions of the joint give typical film thickness values of 20-50 nm. Mixed lubrication in such joints is being achieved by using very smooth surfaces. The roughness levels achieved for these surfaces are 5-10 nm which are feasible with the modern technology available. Such a system can also be a useful tool to study nano tribology with engineering stresses. This example makes use of the conventional theory but is cited as an example where very smooth surfaces can be used to reduce the asperity contact. Besides artificial joints surfaces with nano level roughness find application in magnetic storage devices and micro electro mechanical systems. Most of the engineering systems, however, continue to operate with conventionally manufactured surfaces with roughness ranging from 0.05 to 1.0 μm.

8.4 Pitting in rolling/sliding contacts

In rolling /sliding contacts the repetitive stresses induce fatigue which ultimately leads to pitting. Pitting results in material removal in local areas leading to malfunction of the components. The pits are deep and visible to the naked eye. The term 'spalling' is also used to describe pitting of the catastrophic kind. In some cases there can be micro pitting that may heal itself. In this section the general term 'pitting' is used to describe fatigue induced failure. At this stage it is necessary to distinguish between fatigue wear and pitting. Fatigue wear involves material removal at the asperity scale and is expected to be governed by the cyclic stresses at the asperities. On the other hand pitting involves deeper cracks and visible pit formation. The section first examines why pitting occurs. This is followed by a practical consideration of rolling and rolling/sliding contacts.

8.4.1 Role of contact stresses

The line and point contacts are subject to high stresses even when they operate without asperity contact. First consider pure rolling. The maximum Hertzian stress p_o in such contacts can exceed the yield pressure of the material. As stated in chapter 3 the maximum shear stress is $0.30p_o$ for line contact at a depth of 0.78b while it is 0.31a for a circular contact. These maximum shear stresses act on an oblique plane. There can be local plastic deformation at these depths if the shear stress exceeds 1.67Y. When inclusions are available at these zones, cracks nucleate at these points. The cracks grow with cyclic stressing leading to pits on the surface. When asperity contacts occur there are large pressure fluctuations superimposed on the Hertzian distribution. These local stresses affect the stress distribution in the near surface zone. Below a certain depth the distributions follow the macro distributions obtained on the basis of the smooth surface. The pitting problem now becomes more involved due to the interaction of the stresses at two levels.

In rolling/sliding contacts there is frictional traction at the asperities as well as in the bulk film. The major influence of friction is to induce tensile stresses at the surface. As considered in chapter 3 in a line contact the maximum tensile stress acts at the trailing edge and has a value of $2f p_0$ where p_0 is the maximum Hertzian stress. The tensile stresses act at the asperity level as well as the macro Hertzian contact level and will be higher for the asperities as compared to the macro contact. The influence of tensile stresses is mainly at the surface as these stresses fall rapidly with depth. These stresses lead to the development of cracks at the surface level. Cracks will also be initiated due to shear stresses as discussed in the previous paragraph. With regard to shear stresses an additional factor is to be considered when frictional traction is involved. When friction is involved the point of maximum shear stress moves upwards thereby influencing sub-surface stress distributions. Also the value of maximum shear stress increases with friction. The response of the material to fatigue depends on the complex interaction of the stresses involved making it difficult to predict the fatigue life.

The stress influence on fatigue was considered in terms of a line contact. Similar considerations are applicable to point contacts. With regard to point contacts one difference may be noted. In point contacts tensile stress is involved even in frictionless contact. The maximum tensile stress for this case can be obtained from Eq. 3.6. For typical steel surfaces the value would be $0.133p_0$ and this stress can induce tensile cracks.

It is hoped that the above qualitative consideration will help the reader to appreciate the factors governing the fatigue in rolling/sliding contacts. The asperity contact is now well known to be detrimental to fatigue life. Quantitative theoretical predictions are still not adequate and the present practice involves semi-empirical approaches. The approaches available for rolling element bearings and gears are considered below.

8.4.2 Rolling element bearings

The main approach to fatigue life of these bearings is based on the Lundberg-Palmgren theory [24]. The original equation considered that life was based on the value of maximum sub-surface orthogonal shear stress. Orthogonal shear stress acts parallel to the surface. The maximum orthogonal shear stresses will be somewhat lower than the maximum shear stress considered above and acts at a lower depth. The final equation for bearing life was later expressed in terms of bearing loads and is the most widely used for life estimation. Following the terminology of ISO 281-1990 the relationship is

$$L_{10} = (C/P)^{n_1} L_{c10} \tag{8.1}$$

L_{c10} is the life under a known dynamic load of C. L_{c10} refers to the number of cycles to failure with 90% probability of survival and is equal to 10^6 cycles. Thus C represents the dynamic load that results in a life of 10^6 cycles. The value of n_1 is 3 for ball bearings and 10/3 for roller bearings. P is the equivalent applied load and is calculated on the basis of the given procedure taking into account the radial and axial loads. L_{10} is the fatigue life obtained at the applied load P.

The life of a bearing is affected by the nature of the steel as well as operating conditions. The operating conditions that affect life include temperature, load distribution, and lubrication. It is also necessary to introduce a factor if life has to be estimated for any failure probability other than 90%. This is empirically accommodated by multiplicative factors as given in the standard. Life with any failure probability can now be expressed as

$$L_{na} = a_1 a_2 a_3 L_{10} \tag{8.2}$$

Where a_1, a_2, and a_3 are the adjustment factors for reliability, material properties, and operating conditions. The approaches are empirical and effective selection of adjustment factors is an ongoing debate. Additional factors proposed include those due to dust particles. Any particles that enter the contact will intensify stress effects at local level and so can affect fatigue life.

Fatigue life determination is statistical in nature. Cumulative failures have to be plotted as a function of failure cycles. From such plots and assuming Weibull distribution the life with a given probability of survival is determined. The Weibull distribution function [25] is accepted as the best way to characterise the relation between cumulative failures and life. A large number of failure tests have to be completed to justify the use of Weibull distribution and evaluations are experimentally intensive. Also there is no reason to suppose that the influence of various factors is multiplicative. One factor can influence the other. The more fundamental approach to the problem is based on failure initiating stress proposed by Ioannides et al [26].

Lubricant influence based on Λ ratio has been investigated by Tallian [27] and Skurka [28]. These studies showed clear improvement of fatigue life with film thickness. The observations of Tallian showed that life increased by 1.5 times as Λ increased from 1.0 to 3.0. Skurka found a steeper rise by more than three times for the same variation in Λ. Skurka found that life did not increase with Λ beyond 3.0. However progressive increase was observed by Tallian up to a Λ value of 10.0. Guidance regarding the a_3 factor mentioned above is based on these studies. The lubricants studied were mineral oils without additives.

8.4.3 Gears

Available guidance for pitting life of gears is based on experimental studies and may be considered as a conventional model. As discussed by Dudley [29] the pitting life can be calculated from the following equation

$$L_b/L_a = (W_{ta}/W_{tb})^{n_2} \tag{8.3}$$

L_a is the number of cycles to failure at a given tooth load of W_{ta} while L_b is the number of cycles to failure at any other tooth load of W_{tb}. L_a is the known value at the corresponding tooth load. The value of n_2 is taken to be 3.2 for the boundary regime, 5.3 for mixed regime and 8.4 for full film regime. The values of the exponents once again show the strong influence of Λ ratio on pitting life. Pitting is usually found to occur near the pitch line while the scuffing problems are encountered in the zones of maximum sliding velocity.

It may be appreciated that the prediction of pitting is semi-empirical and can only be obtained with reliable known information on pitting life at a given condition. Also the data are based on case hardened steels and the exponents can be different for different materials. Fundamental approaches based on stress analysis on the lines proposed for rolling element bearings are now being attempted.

8.5 Lubricant-metal interaction effects in fatigue

It is now well recognised that asperity contact through EHD films reduces the fatigue life. As considered above this effect is accommodated in real systems through an adjustment factor for life. There is no adjustment factor to account for chemical additives. This is because the effects are ill understood and this is a research area. Many rolling/sliding contacts use additives. For example gears with Λ ratios less than 2.0 normally use mild EP additives in the oil to protect surfaces from scuffing. A cam and tappet contact in an engine will be lubricated with engine oil containing several additives. It is hence necessary to know whether chemical additives and base oils affect fatigue life. The present section first considers the main findings in laboratory rolling and rolling/sliding testers. The mechanisms are not well understood and the second part discusses the mechanisms from a different perspective.

8.5.1 Laboratory studies

Laboratory simulations to study the lubricant and additive influence have been conducted with disk machines and 4-Ball machines over a long period of time. In fact the early experimental verification of EHD was with disk machines. The tests can be conducted under controlled conditions with proper material control. Also different slide/roll ratios can be attained in the same machine. The present consideration will be mainly confined to the influence of lubricants and additives on fatigue.

In rolling/sliding contacts the slide/roll ratio is defined as

$$\mathrm{SRR} = 2|u_1 - u_2| / (u_1 + u_2) \tag{8.4}$$

Where u_1 and u_2 are the surface velocities of the two rotating disks. Relative sliding velocity is the difference in the surface velocities of the two disks. The average rolling speed is equal to $(u_1 + u_2)/2$. When the surface velocities are equal, and in the same sense, the situation is one of pure rolling and the relative sliding velocity is zero. Normally tests are conducted with surface velocities in the same sense with varying sliding velocities. In some cases the tests may be conducted with opposite peripheral velocities. In such a case, the relative sliding velocity will be the sum of the surface velocities.

Scott [30] and Rounds [31] did extensive work with a rolling 4-Ball tester with high stress levels. In the rolling tester the lower balls are free to rotate thus providing rolling contact. The fatigue failure is assessed on the basis of the vibration level. The machine is cut off when vibration exceeds the pre-selected value. The failure time is relatively low in view of the high stress imposed. Scott reported on the deleterious effects of water content on fatigue. Rounds studied different mineral oils. He also studied the influence of additives on fatigue life. He found that with naphthenic base oil, oleic diamine, oleic alcohol, and oleic acid increased fatigue life at concentrations up to 2%. With higher concentrations the fatigue life decreased. The maximum increase of about four times was observed with diamine followed by oleic alcohol and oleic acid. Similar effects were observed with zinc dialkyl dithiophosphate and sulfurised terpine oil. Chlorinated paraffin reduced fatigue life significantly while TCP had no effect. He also found differences in fatigue life between napthenic and paraffinic oil. Napthenic oils showed higher fatigue life. Steel obtained by different processing routes also affected fatigue life. The influence of additives varied significantly with the nature of refining of the naphthenic oils. Another interesting observation is with regard to pre-formed films formed by heating in additive solutions. A large increase in fatigue life was observed with such pre-treated balls in some cases. The pre-treatment effect depends strongly on the time and temperature used.

A very extensive investigation involving several hundred tests was conducted on behalf of ASME in early seventies and reported in three publications [32,33,34]. The objective of the programme was to understand the influence of material and lubricant chemistry on fatigue life. The tests were conducted in a disk machine at two different SRR values of 3.3% and 30%, which were called slip percentage. Several types of steels and lubricants were used for studies. The contact stresses varied from 1.11 to 4.12 GPa. Large scale regression analysis was done. The data had significant scatter and the regression lines were the best possible fit of this data. The following observations can be made from this work:

1. With paraffinic mineral oils the chemical additives, which showed a beneficial effect in the work of Rounds, decreased the fatigue life in these tests. The base oil used by Rounds was napthenic and this could be one of the reasons for the variation. Oleic acid was also found to reduce fatigue

life significantly. These observations demonstrate that influence of additives is specific to the stress levels, base fluids, and other parameters. ZDTP additive improved fatigue life with increased stress eventually approaching the life for the mineral oil.

2. Fatigue life with low SRR was significantly higher than with high SRR. This may be expected as increased sliding will result in higher temperatures in the contact and reduced film thickness.
3. Commercial pentaerythritol ester with an additive package was similar in performance to the mineral oil at high stress levels.

There are many variables considered including flash temperature, inlet viscosity, pressure coefficient of viscosity, surface finish, and slip ratio. Such a large regression analysis leads to some peculiarities. For example the regression equation resulted in a negative coefficient for viscosity. In other words fatigue life decreased with an increase in viscosity. Several of these issues were discussed and are available with the cited papers. The importance of this work is that generalisations are difficult with regard to additive influence. A directional argument provided in Part III [34] suggested that the additive and base fluid influence is related to a combination of corrosion and protective film formation.

More recently Wang et al [35] conducted rolling fatigue tests in a 4-Ball machine with several different base fluids. They studied eight mineral and synthetic fluids. Experiments were conducted at different stresses that ranged from 8.1 to 9.2 GPa. An interesting feature of the work is the determination of the pressure coefficient of viscosity, α, by experimentation. The main outcome of these studies is that the fatigue life is related to Λ provided proper α value is used in the film thickness calculation. The oils included mineral oils, polybutylene, polyglycols, and a polyol ester. It thus appears that the Λ ratio is the primary variable irrespective of the wide variations in the nature of base fluids. It may be observed that the approach used was to calculate film thickness at a lower temperature of 23°C and a higher temperature of 80°C and explain the trends from the ability to form EHD film. Film thickness in each case at the operating conditions was not calculated. The authors stated that the race temperature varied from 60°C to 90°C. Accurate analysis is only possible when film thickness is calculated at the operating temperatures in each case. Amongst the individual classes the curves relating stress to failure cycles had somewhat different slopes. Two such curves for different fluids with similar viscosity at 100°C are reproduced in Fig. 8.6. The stress-life curve is a familiar representation in the area of fatigue and fracture mechanics. As stress decreases, life increases. When the cumulative number of failures is plotted against number of cycles at a given stress, the L_{10} and L_{50} lives can be obtained from such curves. L_{10} represents 90% probability of survival and will be lower than the L_{50} value. The authors found that ZDTP improved the fatigue life of polyol ester and attributed this to film formation that reduced asperity stresses. These variations in slope show that comparisons on the basis of one stress level alone are inadequate.

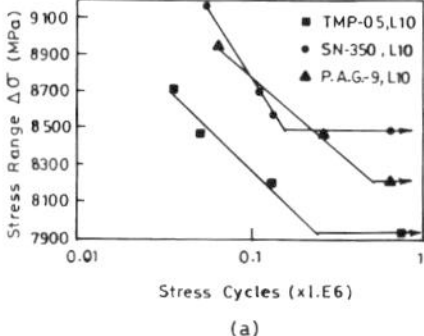

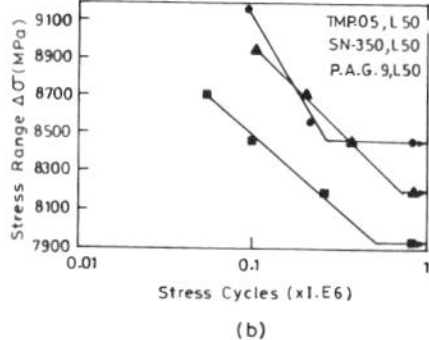

Fig. 8.6. S-N curves with three mineral and synthetic oils with viscosity of 9 cSt at temperature 100°C, S.N.-350 (mineral), TMP-05 (polyol ester) and P.A.G.-9 (polyalkyl glycol): (a) 10% life L_{10}, (b) 50% life L_{50}. (Reproduced from Ref. [35]).

The influence of base fluids can change with temperature. Some fluids can be more reactive inducing crack growth. However the first step is to evaluate the Λ ratio as accurately as possible. Then the chemical effects, if any, can be separated. The main problem is the determination of α for which many laboratories are not equipped.

8.5.2 Observations on mechanisms

The influence of chemical additives is variable. Invoking the possibility of corrosion and film formation as competing mechanisms is to be probed further. Most of the chemical additives tried have a significant beneficial effect on wear in normal boundary lubrication. These include conformal as well as concentrated contacts. Chemical interactions are involved at sliding asperities and the mechanisms involved were discussed in earlier chapters. If increased corrosion decreases fatigue life, it is also expected to increase chemical wear. In reality the additives reduce wear rates by orders of magnitude as compared to the base fluid.

In effect chemical reaction is being invoked simultaneously to explain low wear and low fatigue life. A better explanation is necessary which is attempted below.

Surface cracks are detrimental to fatigue life. These cracks develop at asperity contacts due to tensile stresses and the near surface shear stresses. The near surface shear stress distribution will have a sub-maximum which is higher than the sub-surface maximum value. The influence will be more important with rougher surfaces. Many surface cracks develop, some of which only propagate to critical depth after several stress cycles. The crack grows with each stress cycle in a non-linear manner leading to a pit that may be considered as a local fracture event. These ideas do not take into account the wear process. When there is wear two effects can occur. One is the removal of some of the initiated cracks as they form and the other is the reduction in the depth of the partly grown cracks. Any reduction in crack depth would mean more cycles to pit formation increasing the failure life. Also partial removal of cracks at the initiation stage will also reduce the possibility of development of critical sized cracks. In other words wear contributes to increased fatigue life. Fan et al considered the competition between crack propagation and wear [36]. The emphasis in this paper was on the influence of hardness on such competition. This paper supports the qualitative argument presented here. This argument may be extended now to the role of additives as discussed below.

As wear occurs in sliding contacts it is best to consider a rolling/sliding contact as an example. First consider that an additive is effective and the wear rate is low. Also assume that careful experiments at the given conditions revealed that the fatigue life decreased with the additive in comparison to the base fluid. This is exemplified by the large scale work in rolling/sliding contacts considered in 8.5.1. How can these effects be reconciled? When wear rate is low as with antiwear additives the crack removal rate will be low. Also the crack depths will be less affected. This leads to higher probability of failure. Thus directionally low wear and low fatigue life can be reconciled. Some other additive effects will tend to prolong the fatigue life. The stresses may be more uniformly distributed reducing asperity normal stresses. The tangential stresses also will be reduced due to lower friction. Provided such effects are secondary, the fatigue life should decrease when low wear is involved. This argument may be considered as the mechanical aspect. It does not take into account the influence of the additive itself on crack growth.

The reaction film formation is a rate process. The cracks grow as the film builds. How the additive affects these initial processes is not known and is complex. Unlike the steady state wear in which quasi equilibrium may be envisaged, the initial running-in process is more random. While crack generation can be high at the high spots there is also high wear at these spots and film formation. But pitting life is orders of magnitude higher than the running-in period. It may be reasonable to consider crack growth on the basis of the steady wear zone. In the steady wear zone one can anticipate cases of high crack growth due to chemical interaction. With good antiwear additives strong corrosive reaction does not occur since they form effective barrier films as discussed in chapter 5. Any major corrosive interaction should simultaneously increase the wear and cannot be reconciled with the low wear rates observed with these additives.

Validation of the arguments given so far can, in principle, be done in disk machines by observing the wear rate and crack growth rate with different additives. As crack growth is non-linear, and crack depth itself varies due to wear, the problem is very difficult to handle. Indirect evidence can be gathered if wear rates and fatigue life are obtained over a range of sliding speeds in the boundary regime. Boundary regime will avoid the uncertainties involved in mixed lubrication and it will be easier to interpret the additive effects. It is expected that low wear rates should result in low fatigue life for a given additive if the influence on stresses and friction are secondary. Chemical additives increase fatigue life if 'secondary' effects become 'primary'. As the interest in practice is to obtain increased fatigue life, the 'secondary' influences should become 'primary' influences. In other words positive additive effects on fatigue are possible when crack growth reduces substantially due to low friction and low asperity stresses. For example the reported observation [34] that ZDTP is relatively more effective at higher stress can be due to the nature of film. The film formed at the higher Hertzian stresses may reduce the asperity stresses and friction more effectively. Such an effect in some cases may be related to increased film thickness.

The above argument is valid for conditions in which wear occurs due to relative sliding. In nominally pure rolling wear at the asperity level is expected to be negligible. Also tangential stresses at the asperities will be lower. Though normal stresses at asperities are high the net crack growth will be lower due to the lower tangential stresses. Thus in pure rolling the failure life can be substantially higher. Chemical additives can still react forming films. As there is no wear the influence of chemical additives on fatigue life should be attributed to their influence on crack growth. In the 4-Ball tests considered earlier the stresses used were high and ranged 7.0-9.0 GPa. While such high stresses have the advantage of reducing the failure time, these stresses are unrealistic as there is significant plastic deformation involved in these tests. Chemical reactions can be significant under such high stresses. Two possible additive effects can be envisaged. If reaction film formation is fast enough this will hinder the initial processes of crack generation and growth. In such a case the fatigue life should increase. If the film formation is slow the fatigue life may not be affected. The work due to Rounds [31] showed that in some cases pre-formed reaction films increase fatigue life several fold. It can be argued that these effective films increase life because they hinder the crack growth until they get removed from the surface. This is not possible if the film is ineffective or the additive has corroded the surface during pre-treatment. The influence of such pre-treatments is not known for rolling/sliding contacts. Such studies may provide additional insight into the fatigue mechanisms.

The two effects considered above are mechanical in the sense that the influence of reaction film is only considered in terms of crack growth probability. The third possibility is the influence of the chemical additive on the crack growth itself. This possibility was already considered for rolling/sliding contacts. In the present case the added complication is the short duration of tests in which steady state cannot be assumed. Also reaction film composition will be different for situations involving wear and no-wear. It is hence necessary to apply realistic stress levels to link to practice. More meaningful evaluation for rolling contacts is possible only when stress levels are lower and correspond to the levels used in practice. The evaluations considered in 8.4.2 are usually conducted at realistic stress levels and should be adopted for any meaningful evaluation.

Several factors influence fatigue and the above argument highlights one important aspect that should be taken into account. Several recent experimental and theoretical investigations are being conducted to understand the role of asperity stresses on fatigue. One recent interesting experimental investigation may be cited [37] which provides a state of art consideration of the problem.

8.6 Wear in rolling /sliding contacts

Wear occurs due to relative sliding between surfaces. Pitting life has been extensively studied because of its effect on failure. The wear problems in these contacts are less widely studied. Once the pitting problem is overcome, the component life will be determined on the basis of allowable wear in the system. In mixed lubrication the wear modelling can be approached in terms of the extent of asperity contact through the films. The wear due to the asperity contact is being treated in terms of the available models of boundary lubrication. Limitations of such modelling have already been discussed in the earlier chapters. The theoretical modelling and experimental observations are considered below.

8.6.1 Modelling of contact in mixed lubrication

The contact model of rough surface on a flat due to Greenwood and Williamson has been considered in section 1.3. Firstly consider dry contact. It is usual to solve for the contact parameters in terms of standardised variables. The equations can now be expressed in terms of the standardised variables as follows:

$$n = \eta A_n F_0(h) \tag{8.5}$$

$$A_r = \pi\beta\eta A_n \sigma F_1(h) \tag{8.6}$$

$$L = \frac{4}{3} E\ \beta^{0.5} \eta A_n \sigma^{1.5} F_{1.5}(h) \tag{8.7}$$

where

h = standardized separation given by d/σ

σ = standard deviation of height distribution

$F_n(h) = \int_h^\infty (s-h)^n\ \varphi(s)\,ds$

$\varphi(s)$ = standardized surface height probability density function. For Gaussian surface, it is given by $(1/\sqrt{2\pi})\exp(-s^2/2)$

The contacts made at the asperities can be plastic or elastic. When the extent of deformation at an asperity is such that a maximum Hertzian pressure of 0.6H is reached the contact is considered plastic. If δ_p is the necessary deformation to reach this Hertzian pressure, critical standardised deformation δ_p/σ may be designated as ω_c. All asperities whose deformation is equal to or above this value are considered plastic contacts. This permits separation of elastic and plastic parts of contact as follows:

Number of plastic contacts, $n_p = \eta A_n \int_{h+\omega_c}^\infty \varphi(s)d(s)$ (8.8)

Number of elastic contacts, $n_e = n - n_p$ (8.9)

Real area of plastic contacts, $A_{rp} = \pi\beta\eta A_n \sigma \int_{h+\omega_c}^\infty (s-h)\ \varphi(s)ds$ (8.10)

Real area of elastic contacts, $A_{re} = A_r - A_{rp}$ (8.11)

where

$$\omega_c = (\pi\, 0.6H/2E)^2 \beta$$

The above approach is approximate. In reality there will be a combination of elastic, elastic-plastic, and plastic contacts with varying pressures at different contacts.

These equations have been solved by Rajesh Kumar [38] with a computer programme. An area contact between two steel surfaces with the following parameters may be used to illustrate the results. The parameters represent the equivalent roughness parameters against a smooth rigid disk.

Contact parameters:

$\sigma^* = 0.21$ μm, $\beta^* = 225$ μm, $\eta = 800/\text{mm}^2$, E = 113.19 GPa, H = 6.68 GPa, A_n = 0.1870 mm^2, L = 60 N.

Results:

$n_e = 132$, $n_p = 3$, $A_{re} = 2.80 \times 10^{-2}$ mm^2, $A_{rp} = 0.18 \times 10^{-2}$ mm^2

The extent of plastic contact in this case is low. It will be of interest to calculate the plasticity index as explained in chapter 1. This calculates to be 0.52. As this value is less than 0.6 the contact may be treated as essentially elastic.

The contact problems can be treated in terms of equivalent roughness against a smooth rigid plane only if contacts can be treated as random. As discussed earlier in this chapter when there is conformity at the asperity level the problem has to be treated differently.

Surface roughness in some cases is non-Gaussian. Such problems can be solved by using the appropriate distribution function generated mathematically [39]. A different approach based on the slope of the bearing area curve has been attempted by Rajesh Kumar [38].

The above approaches assume dispersed contact and that stresses at one asperity do not influence the distribution at the other asperities. Also these models are concerned with normal stresses only. How the load distribution varies as the asperity contacts shift does not appear to have been modelled. All that we have at present is a statistical distribution model where the average pressure and the number of asperities in contact are calculated. Such models may be reasonable for steady wear situations. But when dealing with scuffing and running-in, a few large contacts are likely to play a critical role. No effective models are available to treat the problem from this point of view.

The contact in mixed lubrication can be approached in terms of the film thickness. The model considers that any asperity height greater than the film thickness will make a contact. In effect the separation d is replaced by the film thickness h. From the present theories of mixed lubrication in concentrated contacts the problem of contact is more complex and has already been discussed in section 8.2. But a conventional model based on asperity contact ignoring MEHD effects may be considered as the worst possible situation in which maximum contact occurs.

The problem of wear in boundary lubrication has been approached in terms of asperity contact as discussed in chapter 5. Stolarski [40] modelled wear by separating elastic and plastic contact and assigning separate wear coefficients for elastic and plastic contacts. The elastic contacts were considered to wear by fatigue. For plastic contacts the wear coefficient was obtained theoretically. The limitations of such models were also covered in chapter 5.

In mixed lubrication the load is shared by the EHD film and asperity contact. This load sharing affects the film thickness because the EHD film does not carry the total load. Johnson et al [41] proposed the following approximate equation to relate the fluid pressure to the total pressure in such contacts.

$$p_f / p_t = (h_0 / \bar{h})^{6.3} \tag{8.12}$$

where p_f is the fluid pressure, p_t is the total pressure, h_0 is the film thickness based on smooth film theory and $\bar{h}$ is the film thickness to obtain film pressure p_f. This equation shows that the EHD film is stiff and the variation in thickness is very small with variations in fluid pressure. The solution for asperity load sharing can be obtained approximately by neglecting the variation in film thickness. The more accurate procedure would account for the small variations in film thickness. The next issue is the asperity contact model. The complete analysis involves determination of elastic, plastic, and elastic-plastic contacts. Simplifications may be attempted by considering the contact conditions as purely elastic or plastic. These considerations resulted in several models of load sharing in EHD contacts [42,43,44].

8.6.2 Experimental investigations

The major problem with all these models is the lack of experimental verification. Ideally one has to obtain the asperity contact load and film thickness simultaneously. At present this is not possible. Also in view of the present theoretical developments the asperity deformation due to EHD pressure itself has to be taken into account. The above models neglect this influence and the implications are not known. As wear in mixed lubrication is of importance study of wear with different Λ ratios would be useful. Such studies are limited. The work reported by Sastry et al [45] is considered below.

Experiments were conducted in a disk machine with mineral oil at different Λ ratios. The variations were achieved by changing the oil temperature and hence the oil viscosity. EN 31 and low carbon steel disks formed the test pair. The two disks rotated with 10% slip, which introduced relative sliding in the contact. The wear of the mild steel disk was measured as a function of time. The tests were conducted for a duration of eight hours. The relationship between cumulative wear and time was regressed and the expected wear rate between 20 and 21 hours was estimated. This wear rate was called the Wear Factor (WF) and defined as (wear in milligrams/sliding distance (m) x load (N)). The relationship observed is given in Fig.8.7.

The load considered here is the total load. Large influence of Λ ratio on WF can be seen. The following equation describes the relationship.

$$\mathrm{WF} = 8.38\mathrm{x}10^{-4}\,\mathrm{e}^{-3.78\Lambda} \tag{8.13}$$

The correlation coefficient was found to be 0.8, which means the relationship is approximate. Using the models of Tallian [42] for plastic and elastic-plastic contacts the asperity loads were estimated. When WF was plotted in terms of asperity loads, similar curves were obtained but the correlation coefficients were poorer than above.

Empirical relationships of the above nature for different lubricants may be useful for comparison. But any detailed modelling needs an understanding of the wear mechanisms. Adhesive and /or fatigue wear may be involved in such contacts when the lubricant has no chemical additive. Also any modelling of asperity contact in mixed lubrication has a level of uncertainty. Karmakar et al [46] attempted to model wear in rolling/sliding contact under boundary lubrication where the total load is supported by the asperities. It was assumed that the tensile stresses at the asperities governed the fatigue. Modelling of wear needs information with regard to depth of damage and the number of cycles needed for damage. This information was obtained by a new approach by Karmakar et al [47]. In this experimental method cyclic stressing was done on a flat mild steel surface with an EN 31 steel ball. After stressing to different levels the response of the prestressed material to one pass sliding was observed. It was interesting to observe that material removal by fatigue occurred in one pass sliding only when the prestressing exceeded a given number of cycles. The information generated with regard to removal depth, number of cycles, and the calculated tensile stresses formed the basis in modelling wear in rolling/sliding contacts in [46]. The experimental wear rates observed in rolling/sliding contact with an EN 31 and mild steel disk pair showed good correlation with the developed fatigue wear model. Fatigue mechanism was justified in this case on the basis of surface observations as discussed in the cited references. The stress analysis done in this work was approximate and further progress in theoretical modelling is possible. The model is capable of being extended to mixed lubrication if asperity load sharing can be accurately predicted. Depending on the operating conditions and materials the wear mechanism may be governed by adhesion instead of fatigue. In such a case the modelling has to be different.

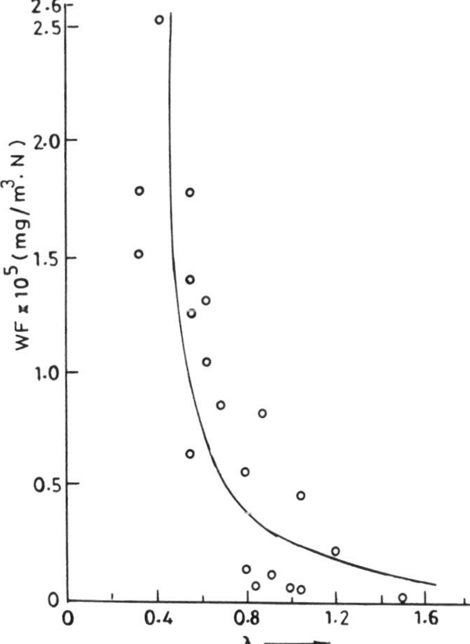

Fig. 8.7. Influence of Λ on wear factor. (Reproduced from Ref. [46] by permission of ASME).

Wear modelling has to be based on the extent of asperity contact and the wear mechanisms operating at the asperities. The issues involved are similar to those already discussed in earlier chapters. One additional factor has to be considered in mixed lubrication. These contacts are subject to high EHD pressures in the whole contact area. Such pressures may additionally influence the bulk reaction in the contact area when chemical additives are involved. As definitive wear models are not available empirical relationships between wear rate and Λ may be developed for different lubricants and additives as discussed above. Such relationships would be useful to compare different formulations. These relationships coupled with surface analysis can also provide a reasonable insight into the mechanisms involved.

References

1. H. Christensen, A theory of mixed lubrication, Proc. Instn. Mech. Engrs., Tribology Group, 186 (1972) 421.
2. H. Christensen and K. Tonder, The hydrodynamic lubrication of rough bearing surfaces of finite width, J. Lub. Tech., ASME, 93 (1971) 324.
3. N. Patir and H. S. Cheng, An average flow model for determining effects of three-dimensional roughness on partial hydrodynamic lubrication, J. Lub. Tech., ASME, 100 (1978) 12.
4. C. C. Kweh, H. P. Evans, and R. W. Snidle, Micro-elastohydrodynamic lubrication of an elliptical contact with transverse and three-dimensional sinusoidal roughness, J. Trib., ASME, 111 (1989) 577.
5. C. H. Venner and W. E. ten Napel, Surface roughness effects in an EHL line contact, J. Trib., ASME, 114 (1992) 616.
6. L. Chang, A deterministic model for line contact partial elastohydrodynamic lubrication, Trib. Int., 28 (1995) 75.
7. N. Fang, L. Chang, and G. L. Johnston, Some insights into micro-EHL pressures, J. Trib., ASME, 121 (1999) 473.
8. X. Jiang, D. Y. Hua, H. S. Cheng, X. Ai, and Si C. Lee, A mixed elastohydrodynamic lubrication model with asperity contact, J. Trib., ASME, 121 (1999) 481.
9. H. A. Spikes, Mixed lubrication-an overview, Lub. Sci., 9 (1997) 221.
10. R. Gohar and A. Cameron, The mapping of elastohydrodynamic contacts, ASLE Trans., 10 (1967) 215.
11. G. J. Johnston, R. Wayte, and H. A. Spikes, The measurement and study of very thin lubricant films in concentrated contacts, Trib. Trans., STLE, 34 (1991) 187.
12. H. A. Spikes, Film-forming additives-Direct and indirect ways to reduce friction, Lub. Sci., 14 (2002) 147.

13. M. Kaneta, T. Sakai, and H. Nishikawa, Effects of surface roughness on point contact EHL, Trib. Trans., STLE, 36 (1993) 605.
14. M. J. Furey, Metallic contact and friction between sliding surfaces, ASLE Trans., 4 (1961) 1.
15. H. Czichos, W. Grimmer, and H. U. Mittmann, Rapid measuring techniques for electrical contact resistance applied to lubricant additive studies, Wear, 40 (1976) 265.
16. A. Sethuramiah, V. P. Chawla and C. Prakash, A new approach to the study of the antiwear behaviour of additives utilizing a metal contact circuit, Wear, 86 (1983) 219.
17. P. M. Lugt, R. W. M. Severt, J. Fogelström, and J. H. Tripp, Influence of surface topography on friction, film breakdown and running-in in the mixed lubrication regime, Proc. Instn Mech. Engrs., Part J, J. Trib., 215 (2001) 519.
18. R. C. Coy and A. Dyson, A rig to simulate the kinematics of the contact between cam and finger follower, Lub. Eng., ASLE, 39 (3) (1983) 143.
19. M. Ram and A. Sethuramiah, Study of lubrication mechanism of two-stroke engine oils using a disc machine, Trib. Int., 17 (1984) 73.
20. M. Ram Tyagi and A. Sethuramiah, Asperity level conformity in partial EHL, Part I: its charecterization, Wear, 197 (1996) 89.
21. M. Ram Tyagi and A. Sethuramiah, Asperity level conformity in partial EHL, Part II: its influence in lubrication, Wear 197 (1996) 98.
22. D. Dowson and Z. M. Jin, Microelastohydrodynamic lubrication of low-elastic-modulus solids on rigid substrates, J. Phys. D: Appl. Phys. 25 (1992) A116.
23. D. Dowson, New joints for the millennium: wear control in total replacement hip joints, Proc. Instn. Mech. Engrs., Part H, Journal of Engineering in Medicine, 215 (2001) 335.
24. G. Lundberg and A. Palmgren, Dynamic capacity of roller bearings, Acta Polytech. Mech. Engng. Ser. 2, 96 (4) (1952).
25. W. Weibull, A statistical representation of fatigue failure in solids, Acta Polytech. Mech. Engng., Ser. 1, 49 (9) 1949.
26. E. Ioannides and T. Harris, A new fatigue life model for rolling bearings, J. Trib., ASME, 107 (1985) 367.
27. T. E. Tallian, On competing failure modes in rolling contact, ASLE Trans., 10 (1967) 418.
28. J. C. Skurka, Elastohydrodynamic lubrication of roller bearings, J. Lub. Tech., ASME, 92 (1970) 281.
29. D. W. Dudley, Gear Wear, in M. B. Peterson and W. O. Winer (eds.), Wear Control Handbook, ASME, New York, 1980, 755-830.
30. D. Scott, Study of the effect of lubricant on pitting failure of balls, Proc. Conf. on Lubrication and Wear, Instn. Mech. Engrs, 1957, 463.
31. F. G. Rounds, Some effects of additives on rolling contact fatigue, ASLE Trans., 10 (1967) 243.

32. M. A. H. Howes, S. Bhattacharyya, F. C. Bock, and N. M. Parikh, Chemical effects of lubrication in contact fatigue, Part I: The test program, Data, and metallurgical observations, J. Lub. Tech., ASME, 98 (1976) 286.
33. S. Bhattacharyya, F. C. Bock, M. A. H. Howes, and N. M. Parikh, Chemical effects of lubrication in fatigue, Part II: The statistical analysis, summary, and conclusions, J. Lub. Tech., ASME, 98 (1976) 299.
34. W. E. Littmann, B. W. Kelley, W. J. Anderson, R. S. Fein, E. E. Klaus, L. B. Sibley, and W. O. Winer, Chemical effects of lubrication in contact fatigue, Part III: Load-life exponent, life scatter, and overall analysis, J. Lub. Tech., ASME, 98 (1976) 308.
35. Y. Wang, J. E. Fernandez, and D. G. Cuervo, Rolling-contact fatigue lives of steel AISI 52100 balls with eight mineral and synthetic lubricants, Wear, 196 (1996) 110.
36. H. Fan, L. M. Keer, W. Cheng, and H. S. Cheng, Competition between fatigue crack propagation and wear, J. Trib., ASME, 115 (1993) 141.
37. D. Nélias, M. L. Dumont, F. Champiot, A. Vincent, D. Girodin, R. Fougères, and L. Flamand, Role of iclusions, surface roughness and operating conditions on rolling contact fatigue, J. Trib., ASME, 121 (1999) 240.
38. Rajesh Kumar, Investigation into the running-in and steady state wear processes, PhD thesis, IIT Delhi, 2002.
39. C.A. Kotwal and B.Bhushan , Contact analysis of non-Gaussian surfaces for minimum static and kinetic friction and wear, Trib. Trans., STLE, 39 (1996) 890.
40. T. A. Stolarski, A system for wear prediction in lubricated sliding contacts, Lub. Sci., 8 (1996) 315.
41. K. L. Johnson, J. A. Greenwood, and S. Y. Poon, A simple theory of asperity contact in elastohydrodynamic lubrication, Wear, 19 (1972) 91.
42. T. E. Tallian, The theory of partial elastohydrodynamic contacts, Wear, 21 (1972) 49.
43. R. A. Thompson, W. Bocchi, A model for asperity load sharing in lubricated contacts, ASLE Trans., 15 (1972) 67.
44. Y. Tsao and K. N. Tong, A model for mixed lubrication, ASLE Trans., 18 (1975) 90.
45. V. R. K. Sastry, D. V. Singh, and A. Sethuramiah, A study of wear mechanisms under partial elastohydrodynamic conditions, Proc. Wear of Materials, ASME, 1987, 245.
46. S. Karmakar, U. R. K. Rao, and A. Sethuramiah, An approach towards fatigue wear modelling, Wear, 198 (1996) 242.
47. S. Karmakar, U. R. K. Rao, and A. Sethuramiah, Characterisation of sliding wear in dynamically stressed material, Wear, B 162 (1993), 1081.

Nomenclature

a_1, a_2, a_3 adjustment factors for reliability, material properties, and operating conditions

A_n	nominal area of contact
A_r	real area of contact
A_{re}	real area of elastic contacts
A_{rp}	real area of plastic contacts
c	radial clearence
C	load at which the life of a rolling contact bearing is10^6 cycles
E	effective modulus of the two bodies
E_1, E_2	elastic modulii of the two bodies, 1 and 2
$F_n(h)$	$= \int_h^{\infty} (s-h)^n \varphi(s)\, ds$
h	standardized separation given by d/σ
$\bar{h}$	film thickness to obtain film pressure of p_f
h_0	film thickness based on smooth film theory
H	hardness
L	load
L_a	known number of cycles to failure at a given tooth load W_{ta}
L_b	number of cycles to failure at any other tooth load W_{tb}
L_{c10}	life with 90% probability under load C and is equal to 10^6 cycles
L_{na}	life for the selected probability with load P
L_{10}	life with 90% probability obtained at the applied load P
L_{50}	life with 50% probability
N	revolutions per second
n	total number of asperity contacts
n_1	exponent
n_2	exponent
n_e	number of elastic contacts
n_p	number of plastic contacts
p	pressure
p_f	fluid pressure
P	equvalent applied load in rolling contact bearing
r	journal radius
rms_z	rms value of the composite profile with conformity
$rms_{1,2}$	rms value of the composite profile based on random contact
R_1, R_2	radii of the bodies 1 and 2
SRR	slide/roll ratio

u	sliding velocity
u_1, u_2	surface velocities of the two disks
WF	wear factor, dimensional
W_{ta}	gear tooth load with known life L_a
W_{tb}	gear tooth load at which life L_b is to be found
Y	yield strength

Greek letters

α	pressure coefficient of viscosity
β	radius of curvature of asperity
δ_p	the necessary deformation to reach the Hertzian pressure of 0.6H
η	absolute viscosity
η	asperity density
$\varphi(s)$	standardized surface height probability density function. For a Gaussian surface, it is given by $(1/\sqrt{2\pi})\exp(-s^2/2)$
Λ	Ratio of film thickness to composite roughness
ν_1, ν_2	Poisson ratios of the two bodies 1 and 2
σ	standard deviation of height distribution
σ^*	composite roughness
σ_1, σ_2	rms roughness of the two surfaces 1 and 2
ω_c	critical standardised deformation given by δ_p/σ

9. Wear in real systems and laboratory rigs

9.1 Introduction

The available knowledge regarding lubricated wear has been discussed in earlier chapters. Several possibilities and limitations exist with regard to the available knowledge. Linking this information to real situations is the purpose of this chapter. As it is not possible to cover all components the chapter mainly considers two different contacts. The selected examples are the piston ring-liner, and the cam follower in engines.

The first section deals with the complexity of wear in real systems and the importance of full-scale tests. This is followed by a consideration of available wear models and practical aspects of running-in. The next section deals with the problem of wear measurement. Newer approaches to measure wear more accurately will be the main consideration of this section. The fourth section deals with strategies for laboratory evaluations in a consolidated manner. This discussion will rely on the material in previous chapters with due consideration of linkage to practice. Thus the vision of the chapter is limited to the gains that can be achieved in the short-term. The long-term needs in the area have already been covered in the earlier chapters. Some repetition from previous chapters was necessary to provide a consolidated presentation.

9.2 The complexity of wear in real systems

Tribological components are sub-assemblies in the overall system. Wear in the components depends on the nature of the lubricant and the extent of boundary contact. In this discussion wear refers to steady state wear. Steady state may be considered as the stable wear after prolonged running. The system operates under varied conditions, which translate into varying wear rates. In many cases there are dust particles in the system despite filtration. These particles additionally contribute to wear. In time the lubricant deteriorates and this is an additional variable. Abrasive wear was not considered in detail in this book. It is very difficult to assess the role of abrasive particles in real systems. During investigation of industrial problems the author has found several cases of high wear due to abrasive particles. The components include turbocharger bearings in locomotive engines and liners of tractor engines operating in dusty atmospheres. In some cases like engines the environment of combustion products also affects wear.

The interest in practice is to minimise component wear and prolong the life. The component wear behaviour may be illustrated by the curve given in Fig. 9.1. The component first goes through a running-in process with high wear rate. This is followed by steady low wear rate. As the wear depth increases beyond a point the wear rate increases significantly and the component has to be replaced. This is related to the increased probability of irregular contact as tolerance limits are approached. Usually the condition also results in increased levels of vibration. When failure rate is plotted in terms of running time a similar curve is obtained that is commonly known as the bathtub curve. The wear levels that can be tolerated depend on the operating conditions and precision needed. For example the performance of a high speed rolling element bearing will be very sensitive to minor levels of surface pitting. Such damage will be tolerable in a low speed operation. Besides prolonged life there can be other reasons for controlling wear. For example the wear in a ring-liner (RL) zone contributes to blow-by and hence emissions from an engine. Wear can also modify contact stresses in comparison to the designed values. The other problem of wear is its non-uniformity. In such cases the life is governed by the zone with the maximum wear depth and not by the overall wear of the component.

The interest in lubricated wear from the point of view of industry was broadly categorised in the beginning of chapter 5. In all cases the important need is to assess wear in real systems. However the wear rates can vary over a range as the component is subject to coupling effects with the rest of the system. As an example it is known that in an automobile the engine liner wear is substantially affected by the number of start-and-stop cycles. One good way to simulate the real situation is by testing the vehicle on test stands where desired operating conditions are imposed. Such tests with prescribed driving cycles are used in the automobile industry. These evaluations can be used to study fuel consumption, oil consumption, and emissions from an engine. The tests can also be done on the engine alone on a test bed. These bench tests will be called sub-assembly tests to distinguish them from the smaller laboratory rig tests. Such tests under prescribed conditions are in extensive use for evaluating lubricants as discussed in chapter 2. Unlike other performance parameters wear evaluation can be done only with very long duration tests of the order of 1000 h for liners. This is because the usual gauging methods used to evaluate wear depth are not sensitive enough. This translates into long test durations. Ring wear and cam follower wear can be measured with shorter duration tests.

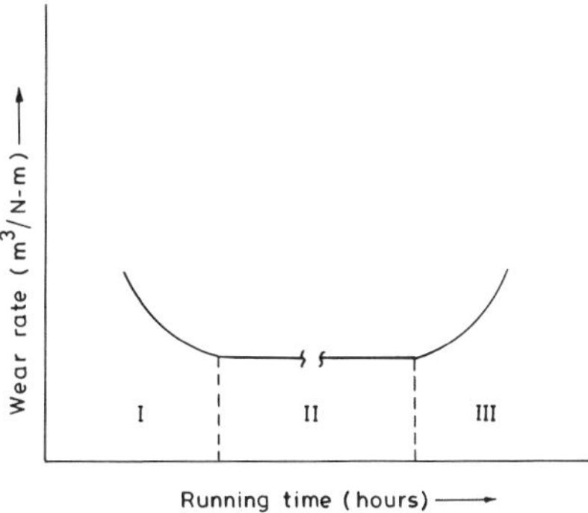

Fig. 9.1. Wear rate variation with running time: (I) Running-in zone, (II) Steady-state zone, and (III) Run-out zone.

Two aspects are of interest with regard to wear tests. The first issue is whether sub-assembly tests will be meaningful. As discussed the wear in real systems will be affected by coupling effects and the environment of the sub-assembly. But in so far as the lubricant-metal interactions go it may be argued that a separate engine test is meaningful for evaluation. Additional variations that occur in real systems are perturbations around this wear behaviour. So distinctions can be made in engine tests with regard to lubricant formulations. The second aspect of importance is the test duration. It is an expensive proposition to conduct very long duration tests. Any effective techniques to measure small amounts of wear will reduce the testing time considerably and much quicker direct evaluations are possible. These developments will be considered later in the chapter.

9.3 Wear in real systems-importance and modelling

This section deals mainly with ring-liner and cam follower wear in engines. These examples are selected because of their industrial importance. Also these are the contacts for which a level of wear modelling is available. First the lubrication of the contacts is considered, followed by a discussion of the importance of wear and models. A brief consideration is then given to other components. The final part considers the running-in in practice and may be treated as an extension of running-in previously considered in chapter 5.

9.3.1 Lubrication of contacts

In IC engines lubricated wear is of importance in the ring-liner (RL) and cam follower (CF) contacts. Wear in main bearings is also a concern when the film thickness is low. One useful book available on engine and vehicle technology [1] may be consulted for detailed understanding of the various components. A specialised book [2] is also available dealing with tribology of the engine components.

In the engine the piston reciprocates between the top dead centre (TDC) and the bottom dead centre (BDC). Piston rings are used for sealing the flow of gases from the combustion chamber. Normally two rings are used for this purpose that are designated as compression rings. In addition an oil control ring is located below the compression rings. This ring scrapes away the excess oil from the surface into the crankcase. A thin oil film on which the rings travel achieves the lubrication between the rings and liner.

The basic lubrication mechanism is hydrodynamic with varying pressures and speeds acting between the ring and the liner as the piston travels. This results in varying film thickness along the stroke. Near the dead centres the film thickness is inadequate to separate the rings from the liner resulting in mixed and boundary lubrication. Minimum thickness is observed few degrees beyond TDC due to squeeze film effect. As the top ring position will be below TDC the maximum wear occurs near the top ring reversal point (TRR). This is illustrated in Fig. 9.2. Though there is reversal at the bottom dead centre also the wear is low due to lower temperatures and loads acting in this zone. In the 4-stroke engine the film thickness also varies between different strokes. Lowest thickness is obtained during the power stroke. The theoretical estimation of film thickness is an active research area for more than 50 years. It is only now that reasonable film thickness predictions are available. The major considerations not only involve the varying gas pressures and temperatures but also the fact that oil flow is restricted due to the influence of one ring on the other. As oil flow continuity is to be maintained the net oil flow decreases and it is now accepted that the lubrication occurs under starved conditions. The major issues involved in this lubrication were reviewed by Ruddy et al [3] in the early eighties and are still pertinent today. Recent attempts are being made to estimate oil consumption, blow-by, and wear with comprehensive simulation of the ring pack [4].

The cam and follower mechanism operates the engine valves. Various configurations are possible [1]. These mechanisms include pivoted valve train systems with finger followers as well as direct acting cam and tappet contacts. The lift is achieved by the lobed part of the non-circular cam. The contact involves high Hertzian stresses and the lubrication is achieved by an EHD mechanism. During a cycle of operation the stresses and sliding speeds vary significantly. A good approach to film thickness calculation is now available based on EHD analysis incorporating squeeze film effects [5].

The main bearings are subject to fluctuating loads and are a special case of hydrodynamic analysis. The mobility method [6] is used for this analysis. More effective methods are also being researched now [7]. Advanced techniques are becoming necessary due to higher loadings on the bearings.

9.3.2 Importance of wear

IC engines are becoming more compact with increasing specific loads. This situation has led to severe operating conditions. Such operating conditions result in higher temperatures and lower film thickness. Further reduction in film thickness is due to the lower viscosity oils being used to decrease hydrodynamic losses. Lower film thickness increases the probability of asperity contact leading to increased wear and associated problems. At the same time increased reliability and longer life is expected from modern engines. Stringent requirements on emissions and oil consumption in the first ten thousand mile of operation also demand low wear rates

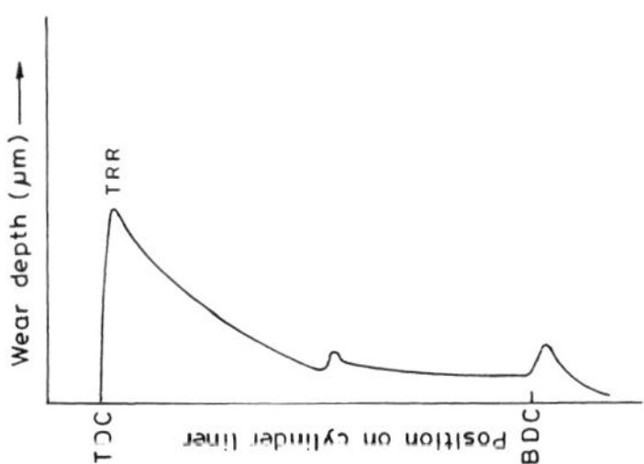

Fig. 9.2. Wear depth variation with position on cylinder liner: (TDC) Top dead centre, (BDC) Bottom dead centre, and (TRR) Top ring reversal point.

in the ring-liner zone as wear intensifies blow-by and oil consumption. Thus wear control is of major importance in the present context. Three possible strategies are possible to control wear.

Wear can be controlled by better chemical additives. As discussed in chapter 4 a large number of new additives are being developed and tested. However as far as engine oils go the main antiwear additive is still ZDTP. Varied structures of ZDTP with different alkyl/aryl groups are possible and this is where modifications are being affected. It is not clear whether any major breakthrough is possible in future. There may be other possibilities when other additives in the formulation provide a synergistic effect. Some new types of friction modifiers may be developed that act synergistically with ZDTP in wear control.

The second approach is the use of wear resistant coatings/materials. Piston ring coatings are already common to provide scuff resistance particularly during running-in. It is expected that the modern technologies available for surface coatings will at least provide partial solutions to the wear problem. Some attempts in the past to use suspended graphite or molybdenum disulphide as solid lubricants were not successful. This is probably due to the inability of these lubricants to displace the adsorbed oil layers on the surface [8]. Any new ideas that allow the particles to coat the surfaces by suitable pre-treatment will be useful. The earlier interest in solid lubricants was friction reduction but any effective coating can reduce wear as well. The possible use of ceramics in some components is another possibility being explored as discussed in chapter 6.

The third possibility is to take advantage of the theoretical ideas available in mixed lubrication. As discussed in chapter 8 modified roughness patterns can assist in forming thicker films. Recently Priest and Taylor [9] discussed these possibilities. They have given examples of surface modification for cam and follower contacts as well as micro grooving to improve film thickness in main bearings. One problem in applying the theoretical ideas is the significant modification that occurs in roughness due to the running-in. In effect what is needed is the required pattern beyond running-in. Hence, implementing ideas in practice is more involved. Another aspect of roughness is the possibility of asperity level conformity as discussed in section 8.3.3. Coy [10] has given an example of such conformity in a cam and pivoted follower contact. There is no detailed information on this aspect with different cam and follower mechanisms. Investigation of this aspect should be useful. Such conformity can significantly reduce asperity contact for a given thickness and is important in the present context.

The engineering solution of the problem is likely to be a combination of the above approaches. The current film thicknesses observed in modern engines are already low. As reported in [9] the minimum film thickness observed in the ring-liner zone can be as low as 0.2 μm for modern gasoline engines. The cam and follower operates with minimum thickness of 0.1μm. Even in main bearings that rely on hydrodynamic lubrication the minimum thickness is around 1.0 μm.

9.3.3 Modelling of wear

The available methods to calculate film thickness provide a good estimation of film thickness. Wear modelling on the other hand lags behind and is empirical. As discussed in earlier chapters wear modelling cannot be done on the basis of fundamental considerations even in controlled laboratory experiments. The present approach to modelling is considered below.

The initial approach to wear in ring-liner contact was based on a limiting film thickness [11]. In this modelling wear was considered to occur only in those zones where film thickness was below this level. Wear in these zones was modelled on the basis of Archard's equation assuming a constant wear coefficient. The present wear model as discussed by Priest and Taylor [9] is based on a modification of the above idea. In this model a constant boundary wear factor k_b is assumed for Λ ratio ≤ 0.5. Wear volume for the boundary zone is expressed as

$$V = k_b W l \tag{9.1}$$

where V is the wear volume, k_b is the wear factor, W is the load, and l is the sliding distance

It may be noted that the wear coefficient has the units of specific wear rate. It was obtained from laboratory tests the details of which were not given. The terms 'boundary wear' and 'boundary lubrication' are used in a broader sense throughout this chapter and in reality refer to chemical wear in most cases.

For the mixed lubrication zone the wear coefficient is assumed to be proportional to Λ. The linear relation is obtained by taking k_b is zero for $\Lambda = 4.0$ and the maximum value at $\Lambda = 0.5$. The wear factor at any Λ value $k_b(\Lambda)$ is obtained from this relation. Wear volume and hence the wear depth can be estimated from this equation at different liner locations. The ring wear can also be obtained by integrating the changing wear volume in each stroke. It is claimed that this approach reasonably predicts the experimentally observed wear patterns.

There are two problems with such modelling. The first problem is the value of k_b in boundary lubrication. There is no doubt that k_b varies with temperature. Both fundamental considerations and empirical modelling in a reciprocating rig reported in section 5.4.1 show that the wear will be influenced by temperature. Over a temperature range of 50° to 150°C the wear rates for engine oil decreased by about three times. The material combination was an EN 31 ball sliding against an EN 31 flat. As fundamental modelling is not possible the possibility of empirical modelling may be considered. This involves the recognition that k_b is a function of temperature and is proportional to T_c^a where 'a' is a negative exponent and T_c is the contact temperature in °C. The value of 'a' has to be derived empirically from the experimental data. It may be considered that the temperature influence is the same for both k_b and $k_b(\Lambda)$. Implicit assumption here is that wear coefficient is independent of load and speed, which may be acceptable. Surface temperature rise in the contact due to friction is not known and its influence also needs a theoretical analysis. Here again the relations may be tried on the basis of liner temperature only as a first approximation. It may be also noted that the reaction film is a 'mixed' film since the film on the moving rings with varying temperature is changing and interacting with the liner film. This is unlike a laboratory experiment where film behaviour can be characterised at one operating temperature. Under such a situation empirical consideration of the temperature influence appears to be the only possibility.

The k_b value should be derivable from the experimentally observed wear at the TDC. The values of k_b observed for different formulations will be very useful reference data for laboratory rig tests.

Another problem with the available model is the assumption that wear reduces linearly with Λ. The experimental work reported in section 8.6.2 showed significant non-linearity in wear under partial EHD conditions. It appears this assumption has to be experimentally verified for the ring-liner contact that involves hydrodynamic lubrication. Probably evaluations in a motored engine using only mineral oil may be useful for this purpose. As film thickness can be calculated along the stroke the variations in liner wear can be related to the Λ ratio. The nature of the relationship can then be used with confidence.

Cam and follower contacts operate in the EHD regime and the cyclic variations in film thickness can be reasonably predicted. In this case again wear is modelled by the same procedure as in the case of ring-liner contact. Reasonable prediction of the wear pattern is observed in this case also as reported by Bell and Colgan [12]. They used a boundary wear factor value of 1×10^{-17} m^3/mN based on earlier laboratory tests. The problems involved with this modelling are similar to those discussed for ring-liner contacts. It may be possible in this case to model wear as a function of Λ with a disc machine with appropriate material combination. The appropriate method has to be researched. This will enable firming up at least one aspect of the wear model. In so far as the influence of temperature on k_b goes, it may have to be accommodated empirically.

Some engine tests are available as given in Table 2.2 of chapter 2 to specify wear limits. Such specifications already exercise a degree of control on wear. Some of these tests have the potential of being adapted for comparative wear studies.

The above discussion shows that fundamental research should now focus on real systems. The gaps involved are major and the possible strategies were discussed in chapter 5. In the short-term, improvement in the existing methods should be sought as discussed above. It is also necessary to explore modern techniques like FFM. For example scratch tests on the worn surfaces may provide an indication of the wear resistance of the films. Such information will be valuable in realistic wear modelling. Successful use of tribology for industrial problems is of importance for its widespread acceptance and should be actively pursued.

9.3.4 Wear in other components

The author has so far considered two typical contacts with regard to wear modelling. The contacts considered are industrially important and there is work reported on wear modelling. In many other systems like gears, hydraulic systems, roll wear in cold and hot rolling the modelling is not even at the level considered in the above contacts. It is expected that the existing models for ring-liner contact can be extended to reciprocating contacts leading to a reasonable prediction capability. In many of these contacts the problem of wear is known and selection of lubricants is based on experience. For example a synthetic lubricant may perform better than emulsions in high speed cold rolling and will be the preferred lubricant subject to cost considerations. In high-speed gears there is more interest in preventing scuffing. This is evidenced by large amount of effort to model scuffing. More recently there have been models published with regard to gear wear [13]. These models take into account the changing contact stresses due to wear and integrate the wear in each cycle incrementally. However the wear prediction is simply based on a constant wear coefficient. The value of wear coefficient and its variation is the crux of the problem in wear modelling as discussed above. Some other wear equations relate wear life to operating conditions empirically. One example for life of cutting tool [14] may be cited. The relationship is expressed as

$$T_t v^{1/h} t^{1/m} b^{1/p} = \text{constant} \tag{9.2}$$

where T_t is tool life, t is feed rate, v is the cutting speed, and b is depth of cut and n, m, p are the exponents that satisfy the equation.

Such equations are general in nature and provide some prediction capabilities as a function of operating conditions for a given system. It may be noted there is no explicit parameter for the lubricant. Any variations due to lubricant cannot be predicted from this equation and can only be assessed from experiments.

Equations to predict wear quantitatively are not well developed. But the efforts to reduce wear with different strategies will lead to lower wear rates in future. This is of importance irrespective of the capability to model wear precisely. The parallel development needed is the ability to classify wear in laboratory rigs in the same order as observed in the real contact.

9.3.5 Running-in in practice

Running-in of the components has to be done in the real system and its effectiveness has to be characterised. The major research reported in this area is with regard to ring-liner contacts. The nature of honed surface roughness for a liner is given in Fig 9.3. The plateaux are smoother zones obtained in the two-stage honing process that first involves rough honing with a cross hatched pattern. This is followed by fine grinding of the rough surface resulting in smoother plateaux. Different levels of finer roughness at the plateaux can be obtained by varying the manufacturing process. Smoother plateaux were considered to be closer to the run-in surfaces and hence need lower time for running-in. On the other hand smoother surface plateaux are more prone to scuffing during increased loading [15]. Some laboratory studies recently reported also seem to support the idea [16]. Thus some optimum roughness is to be sought for the plateaux. Several approaches to characterise the initial surface roughness have been proposed. Completion of the running-in has been characterised by the expected changes in these parameters [17]. The important issue is to know what is the best procedure to adopt for the quickest running-in of a given engine. There are no effective answers to this question and the available procedures are empirical as stated in chapter 5. From the point of view of practice another aspect to be investigated is the influence of running-in and initial roughness on steady state wear. This has implications to the liner life. Laboratory studies reported in chapter 5 indicated that initial roughness influences steady state wear though final roughness may be same during the steady state. The material combination in this case involved EN 31 steels. This may be related to the mechanical properties of the films that develop slowly with time of running. Thus steady state wear is also a parameter to be taken into account in selecting initial roughness and the running-in procedure.

One practical approach is the use of surface coatings on rings and in some cases on the liner. Chromium and molybdenum coatings are widely used for piston rings. Gas nitriding is also being used for surface modification. Even with such coatings the effective running-in procedure has to be empirically developed though the time needed may be shorter.

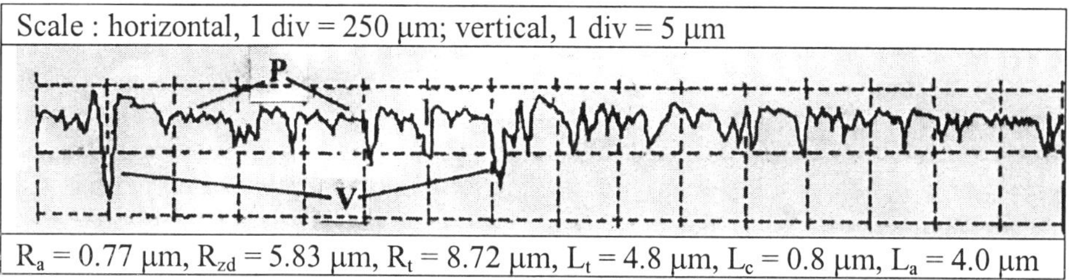

Fig. 9.3 An example of honed surface roughness profile (P) Plateaued zone (V) Valleys. (Reproduced from Ref. [22])

In the past fine abrasives were used to assist in the quicker running-in by removing asperities faster. Another approach tried was the soluble chromium salts that eventually formed oxides that acted as fine abrasive. Such procedures have the problem of using a separate oil during running-in as well as the complication of the abrasive particles that remain in the system. While some commercial oils are available for the purpose the extent of their application is not known. One additional possibility is to develop protective oxide films on the surface during final stages of manufacture. The author has not come across any research in this direction.

The associated problem is to assess surfaces and coatings in the laboratory. Large amount of work has been reported in this area with a wide variety of test configurations [18,19]. They have similar problems in interpretation as have been discussed for antiwear and EP additives in chapter 7. Here again mapping is necessary over a wider range of operating conditions.

Running-in is also of importance for other components like gears and bearings. While some guidelines are reported the procedures are at best semi-empirical.

Running-in can change friction coefficient and contact temperature and these parameters can be used as indirect indicators of the running-in process. Blau [20] has given an elaborate consideration of these possibilities that are not considered here.

The problem of load carrying capacity has already been considered in earlier chapters. Current laboratory tests and their limitations were already discussed. The guidelines for practice mainly relate to contact temperatures and are mainly applicable to gears. For many other systems experience seems to be the only answer despite some guidance from laboratory EP tests.

9.4 The problem of wear measurement

Any assessment of wear by conventional techniques like gauging of liner wear have limitations. There can easily be an error of ± 1.0 μm in such measurements. Hence for assessment of wear several hundred hours of testing is needed as wear in many engines will be of the order of 1.0 μm per hundred hours in the zone of highest wear. Any comparisons with laboratory wear tests become difficult, as it is expensive to run such long duration tests. This is also true with many other systems like gears in which the wear rates are low. Accurate wear measurements in short duration tests will help in quicker evaluation of lubricants. It will assist in faster comparison of wear between laboratory and large-scale tests. There is growing interest in such evaluations and the three main reported approaches are considered below. The first two approaches are related to the liner wear measurement and are discussed under this category. The third approach based on roughness trace analysis is separately considered as it is being used for measuring very low wear. All evaluations depend on the comparison of roughness parameters but differ in the methodology and scale of observation.

9.4.1 Removable insert method

Henein et al [21] have reported an interesting new approach to wear measurement. They have inserted a wear probe in the liner, which was removable. The probe was machined carefully so that it was flush with the liner. The probe was so located that the top ring was centred on it at TDC position. The top quarter of the probe does not contact the ring and was used as a reference surface. The probe was manufactured so that its position in the cylinder liner was reproduced accurately whenever it was reinstalled. An area of 2.2 x 3.5 mm was selected for scanning. The 3-D and 2-D roughness of the worn surfaces was obtained by a laser stylus meter. The system with the associated software was also equipped to measure the total volume above a given reference. The difference in volume after due correction for any change in reference level gave the wear volume of the scanned portion. The nominal wear depth was obtained on the basis of the scanned area. The method was used to study wear in five unequal steps with a total time of 3 h. The important finding was that major wear occurred in the first hour for the selected operating conditions of the engine. The average wear rate in the first one hour was 12 times that during second and third hours. Several roughness parameters obtained before and after wear were also compared.

The above method should be applicable for long duration tests as well depending on the selected reference level. The method is surely of value but the limitation is the availability of the techniques used. As the authors' pointed out this can be a valuable tool to quickly assess the running-in process for different conditions and to ensure that the initial running-in is effectively completed. In principle the method should be applicable for wear evaluation in any reciprocating machine.

9.4.2 Bearing area method

The liner is usually honed and the valleys may typically range from 4-10 μm. Such honing is considered advantageous for lubricant retention. The roughness trace of a typical honed surface was illustrated in Fig. 9.3.

Since the bearing area curve was not considered earlier it is explained here. In a 2-D profile trace several parallel lines may be drawn. Each line cuts through solid and empty space. The percentage of the length that passes through solid at each depth can be plotted as a function of depth. This is normally called a bearing area curve. This is illustrated in Fig. 9.4. This curve will now be referred to as a bearing length curve though it is conventionally called bearing area curve. The percentage of bearing length increases with depth.

The engine wear in short duration tests of the order of 100 to 200 h is low and will occur within the honed roughness. The idea that has been used for estimation is to have an internal reference that will be the same for worn and the unworn liners. The approach used by Rajesh Kumar et al [22] considered the 90% point of the bearing length curves to be the same before and after wear. The bearing length curve after wear was superimposed on the original curve with common 90% point. The area between the two curves provided the worn area over the assessment length. At each measurement zone the volume was obtained by considering the area to be representative over a circumferential distance corresponding to one degree. The wear depth was obtained on the basis of difference at 0% bearing length with some modification. The procedure to find the worn area is illustrated in Fig. 9.5.

The above procedure was used to assess wear in 200 h endurance tests. The tests were conducted on a two-cylinder, four-stroke, air-cooled tractor engine of 34 BHP (25 kW). Effective comparison between liners subjected to two different running-in procedures was possible by this procedure. The radial wear depth ranged from 0.7 to about 3.4 μm depending on the distance from TDC. The air-cooled engine did not show the expected uniform decrease along the stroke length. The possibilities and limitations of the method were discussed in the cited reference. The procedure is based on nominal relocation at different distances and 2-D profiles. Estimation based on 3-D profilometry and better relocation will improve the utility of the technique.

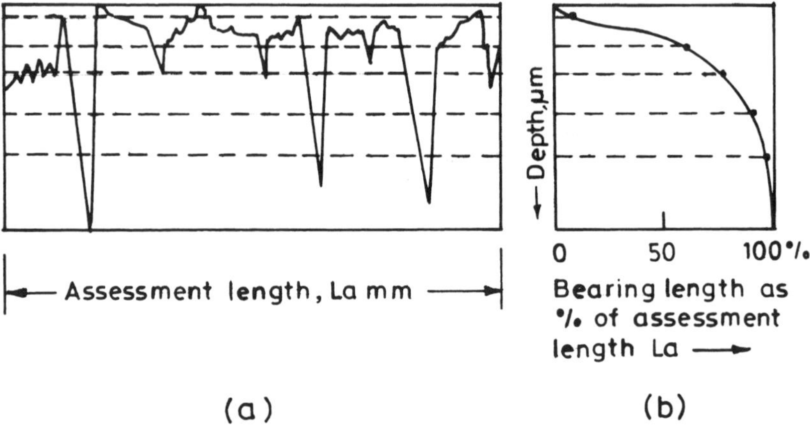

Fig. 9.4. Schematic diagram of roughness profile with corresponding bearing length curve (a) Roughness profile (b) Bearing length curve.

9.4.3 Precision techniques for wear measurement

As discussed above any wear evaluation needs a reference. For honed surfaces with deep valleys advantage can be taken of the fact that the deeper part of the surface can be used as a reference. For surfaces without honing, and low roughness wear can be measured with reference to some deeper unchanged part like a valley. Such measurements will be meaningful for very small depths only. One technique used

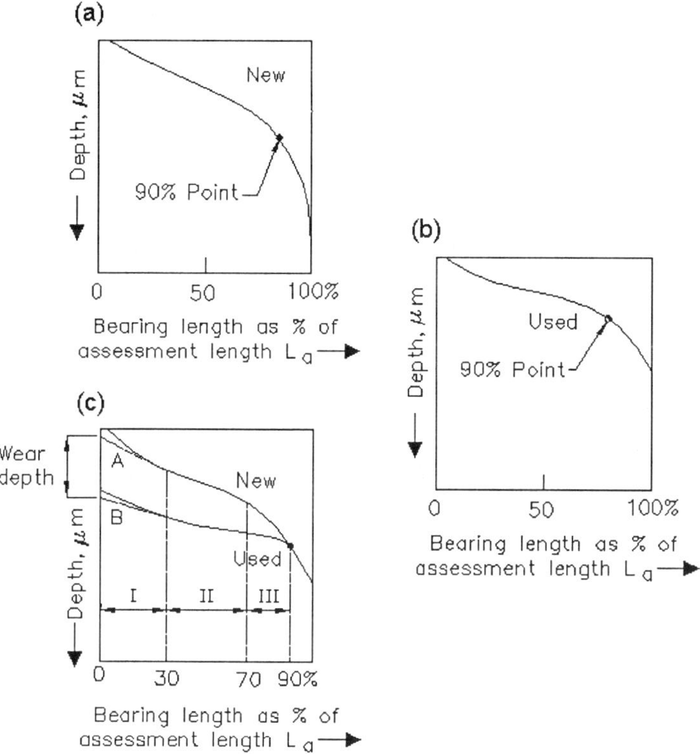

Fig. 9.5. Methodology for wear estimation using bearing length curve: (a) and (b) Bearing length curves of new and used liners, and(c) Wear estimation after matching 90% point. (Reproduced from Ref. [22])

involved AFM techniques [23]. The advantage of this technique is the generation of 3-D topography before and after wear. Matching can be done on the basis of relatively deeper unaffected portions. Point to point height variations can be compared. The wear volume can also be obtained by subtracting bearing volumes obtained before and after test. Another approach reported [24] involved accurate relocation of the surface to the extent possible by specially developed methods used for precision machine tools. The roughness measurements were done on a normal stylus instrument. The relocation was further improved by software techniques. Some techniques based on image processing have also been reported [25]. All these methods lead to good assessment of the small wear that occurs on the surface as well as its distribution. Tests are also possible by using a fine scratch as a reference. The change in scratch depth itself can provide a rough indication of relative wear depth. A recent paper used this technique to assess seal wear [26]. The authors were able to put in evidence the variations in wear depth due to thermoelastic deformations in a carbon graphite seal. It is expected that wear measurement techniques will advance significantly in the near future. The problem that has to be resolved is the scale of observation that is meaningful for specific applications.

9.5 Strategies for laboratory evaluation

Sub-assembly tests are the best way to evaluate lubricated wear. In principle comparison between formulations as well as adequacy for new systems is possible by these tests. The techniques being developed now for measurement as discussed above will facilitate such evaluations. But even with reduction in test time using such tests on a routine basis to evaluate formulations is not a practical solution. This is because reasonable test duration is necessary for comparisons. Strategies are obviously necessary to make laboratory rig evaluations more meaningful and the present section addresses these problems. The emphasis is on ideas that provide improvement in the short-term. Some aspects considered in earlier chapters are repeated to provide consolidated treatment. The issues related to rig selection and operating conditions, lubrication regime, and wear mapping are first considered. The final part considers the linkage of laboratory tests to practice. In this section 'lubricant' and 'formulation' are used interchangeably. Antiwear additives are only referred to as additives unless otherwise qualified.

9.5.1 Selection of test rig and operating conditions

Several rigs are available for testing. The rig selection should preferably be based on the nature of contact. Thus it is good to use a reciprocating tester with area contact to study ring-liner wear. For gears as well as cam and followers disk machines are the obvious choice. The test material combination should be the same as the real system. These considerations are elementary but often ignored in lubricant evaluation.

Any test rig that may or may not simulate the stresses and materials will still be capable of ranking lubricants. The problem with such ranking is that influence of contact stresses and materials are changed. Correlations between different rigs over a wide range of operating conditions is necessary for this purpose which is not available. So the best strategy is to at least simulate what can be simulated.

As has been emphasised in chapter 7 any ranking based on one test condition is totally inadequate. It is essential to develop wear maps on the lines suggested in chapter 5. The operating range should preferably cover the conditions in practice. Some operating conditions like velocity cannot be easily simulated. For example the average sliding speed in a real ring-liner contact can easily exceed several m/s while many reciprocating testers operate at an average velocity of less than 1 m/s. In this specific case low speeds are meaningful with regard to wear near the TDC where speeds are low. The temperature range that has a strong influence on wear can be easily simulated. The required stress ranges can also be normally simulated.

All area contacts present problems with regard to alignment of the test pieces. Careful design is necessary to ensure alignment. Even with good design it may be necessary to do pre-running to ensure alignment. With concentrated contact like a ball-on-disk the problem can be avoided. As the ball wears the contact changes to area contact and the contact can simulate area contact. The area, and hence the apparent pressure, keeps varying as the test progresses. The widely used 4-Ball wear tester is an example of varying area contact. Material simulation is possible with a hemispherical contact sliding against a disk. The 4-Ball machine also can accommodate different materials and configurations though the changes are more difficult.

New and innovative ideas are possible for laboratory rigs. The tests need not be confined to known standard test rigs. The advantage with standard rigs is the known levels of repeatability and the levels of precision reached in manufacture. Also these rigs are widely available and comparisons between laboratories are easier. One possible innovation is the careful modification of the existing test procedure to suit requirements. A modified test procedure say, in a 4-Ball EP tester, may correlate with an FZG rig in EP performance. The difference in EP action between these rigs was discussed in section 7.4.3. Such considerations can be used to develop correlations between rigs. The idea here is not to emphasise the correlations between rigs per se. The important possibility of wide variation in conditions should be better exploited in the existing rigs.

The need for new rigs is particularly relevant when the operating conditions in real system cannot be simulated in the existing rigs. Some examples discussed in chapter 5 were with regard to a hypoid rear-axle and high-speed cold rolling. In many such cases involving high severity, reliability is more important though better lubrication is expected to reduce the wear.

9.5.2 Lubrication regime for testing

The regime to be used should be boundary lubrication. Any partial hydrodynamic effects complicate the interpretation of wear data. In wear tests particularly at high speeds some hydrodynamic effects can occur. Such effects can be indirectly assessed on the basis of friction coefficient as well as variation in cumulative wear with time as discussed in chapter 7. Real components operate in mixed and boundary conditions depending on the nature of contact and operating conditions. But in so far as distinction between the additives is concerned, the wear rates should depend on the boundary contact zone. This is valid provided the mixed lubrication conditions are similar for the different additives. This assumption is reasonable if tests in the sub-system are conducted under the same conditions. One complication that can arise even in such testing is the influence of the additive on roughness. Suppose one additive has a significant smoothening effect on the surface. In such a case the mixed lubrication effects change and this can influence comparison of wear rates. These effects may or may not be significant and need to be assessed on the basis of the final roughness of worn surfaces. Another complication is related to asperity stresses. In mixed lubrication the load is distributed between asperity contacts and the fluid film and the contacts will be essentially elastic. If the total load is supported by boundary contact some asperity contacts can reach plastic deformation. The effect will not be significant if the contacts are essentially elastic in both cases. Usually in laboratory wear testing where area contacts are involved the contacts are expected to be mainly elastic in long duration tests.

The above remarks apply to machines that work with area contact. In the case of disk machines mixed lubrication will be common because of significant EHD effects. One way of assuring boundary regime is to work with Λ ratio less than 0.5. Such a requirement will restrict the choice of operating conditions. This problem can be overcome if asperity contact levels in mixed lubrication can be modelled. This is difficult as discussed in chapter 8. One possible way is to develop empirical relationships between Λ and wear rate as discussed in section 8.6.2. Here again the methodology has to be researched. In cases where there is a level of asperity conformity, composite roughness has to be evaluated differently as discussed in section 8.3.3.

9.5.3 Development of wear maps

The first step is to define wear rate. Theoretically the run-in process continues over a long period of time and steady wear rate is what is obtained after infinite time. Modelling wear rate on this basis was discussed in section 5.4.1. From the available computer programmes steady wear rate can be obtained. In practice the change in wear rate becomes negligible beyond an adequate running time. The practical situation with regard to 4-Ball evaluation was discussed in chapter 7. The suggested approach was to evaluate wear over a relatively long duration with step-wise determination of cumulative wear volume. The information can be used to obtain wear rate by regression analysis. Though this is by definition not steady state wear rate, it is expected this value is meaningful for comparison purposes. Quantitative change between any two lubricants will be different for the two cases. But it is reasonable to assume that ranking in relatively long duration tests is adequate from a practical point of view. The actual test duration needed depends on the operating conditions. The best approach can be to decide the running time on the basis of selected total sliding distance. As discussed in chapter 7 the author considers the present classification of wear with short duration of 30 minutes or so to be inadequate and the testing strategy must change. It may be appreciated that 'long duration' is only qualitatively expressed and has to be based on exploratory testing work. A quantitative approach may be possible on the basis of rate of change in wear rate with time.

The necessity for wear maps has been discussed in chapter 5. Such mapping as shown in Fig. 5.7 has clear relevance to practice. The question with regard to practice is the extent of mapping that is necessary. For this purpose it is necessary to appreciate the time requirements. Ideally in a given tester the variables to be considered are the load, temperature, sliding speed, and surface roughness. For lubricant evaluation it is possible to fix the initial roughness of the test pieces, which leaves only three variables. If the variables are tested at three levels each, a full factorial design will involve 27 test combinations. Assuming each test is repeated at least once the total number of experiments will be 54. With each test run lasting at least for three hours the total test time for evaluation of one lubricant will be about 200 h. The testing can be reduced for example by selecting one-third fractional design that will involve a total testing time of nearly 70 h. Such partial designs will be less accurate as compared to full designs. Once the data is available useful empirical relations can be developed as discussed in chapter 5.

The wear maps discussed above are illustrative only. The variables may be tested at more than three levels in which case the mapping will take more time. In some cases different material combinations and roughness effects may be important variables needed for study. It may be appreciated that specific guidelines cannot be evolved for each case and the investigator has to make his choices. The author considers that the first step in screening can be based on a study of the influence of temperature only. In this approach the wear rate is studied with a fixed load and speed at different temperatures. If possible the load and speed conditions should correspond to practice. The temperature range selected must include the operating temperatures in the real system. Even in this case development of a relationship can mean at least 20-25 h of experimentation for each additive assuming five different temperatures are selected and some repeat tests are done. The total contact temperature that includes the surface temperature rise should be the governing temperature for evaluation. The surface temperature rise calculation may be based on the geometric contact area as discussed in chapter 1. While asperity temperatures can also be estimated as discussed in chapter 3 they are less certain as asperity contact sizes are not known. In effect the overall contact temperature may be treated as a thermal parameter that reasonably describes the overall influence on wear. The speed and load effects are not known and it cannot be assumed that their influence will be related to the change in contact temperature alone. This initial screening will be helpful in identifying the candidate additives. The selected additives may then be studied with more extensive mapping.

The wear maps have the advantage of providing information under different conditions. As an example a formulation for gear oils may operate under different conditions of load and speed. The ideal candidate will be the formulation with minimum wear but also least possible variation over the range of operating conditions. Assessment of formulations from this point of view is not possible without mapping. In some cases the formulators already know from experience several aspects of the performance in real systems. With such experience development of maps can be restricted to the required zones only where experience is not available. The wear map approach quantifies performance unlike experience that is generally qualitative.

9.5.4 Linkage to practice

The discussion so far has been related to the best approach to compare formulations in a given test machine. It is important to know how they are related to the wear rates in the large-scale tests. Here the comparison is discussed only in terms of sub-assembly tests for reasons already stated. The wear in a sub-assembly depends on the extent of asperity contact through mixed films and the applicable wear coefficient in the boundary contact. Any prediction of Λ ratio needs estimation of film thickness as a first step. As discussed earlier the film thickness can be effectively estimated for complex geometries involving ring-liner and cam follower in engines. These developments are capable of being extended to many industrial situations involving reciprocating area contacts like compressors. Also for EHD contacts film thickness can be estimated with EHD theory. Thus as a starting point it is assumed that the information regarding Λ values is available. The two areas of uncertainty are now the influence of Λ on wear rate and the applicable boundary wear coefficient. The wear coefficient may be expressed in units of specific wear rate or as a non-dimensional coefficient as convenient. The problems involved are already discussed in section 9.3.3. The role of laboratory evaluations will now be discussed in terms of bridging the gaps. The extent to which laboratory evaluations can be used in choosing alternate formulations will be considered next.

The ideal approach is first considered. To obtain information on laboratory machines wear mapping has to be done under boundary conditions over a range of operating conditions that will include the operating conditions of the sub-assembly. These tests may be conducted with area contact or line contact as applicable. The empirical relations developed will provide the required boundary wear coefficients at different operating conditions. Such information can be generated for various additive formulations. The same formulations may then be evaluated with known variations in Λ ratios. Varying Λ ratios in EHD can be generated in disk machines. But when partial hydrodynamic effects are involved the influence of Λ ratio can be different and suitable rigs may have to be devised. For example a parallel thrust pad with varying levels of hydrostatic lubrication can be used for such a purpose. A comparison of the wear maps in the two cases should be capable of providing relationships combining mixed lubrication as well as boundary effects. Such development is not straightforward and needs substantial research. Also comparison has to be done with the wear information generated in the sub-assembly to substantiate the models. The effort needed can only be justified in terms of the cost/benefit ratio. The recent interest in engine wear may justify such an effort. The obvious advantage of the suggested approach is the generation of information to directly predict wear rate in the real operating systems.

Partial implementation of the above ideas may be considered as the next best alternative. The development of wear maps under boundary conditions alone can be very useful in ranking formulations. Mixed lubrication effects can decrease the asperity contact as well as the stresses at the asperities. Both these effects can only reduce the boundary wear. Though real wear cannot be predicted in terms of these effects the relative wear is expected to be governed by the basic boundary wear behaviour of additives. It is hence reasonable to rank formulations on the basis of wear behaviour in boundary contact conditions. Even here some problems have to be envisaged regarding development of wear maps and the corresponding wear equations. While contact stress and temperature can be simulated there can be difficulties in simulating speed effects. Here speed effect refers to the influence of speed on boundary lubrication. Any extrapolation of the empirical equation to high speed conditions needs a judgment based on observed wear behaviour as a function of speed. For example if speed effects on wear coefficient are small it is reasonable to ignore this effect. Other problems that can arise include the inability to simulate contact conditions or materials. These issues were already discussed and are not considered here. Yet another practical problem can be the inability to avoid mixed lubrication under certain operating conditions. The relationships in such cases have to be restricted to the cases involving boundary lubrication. Despite these limitations the information generated will provide the ability to rank different formulations with regard to wear under different operating conditions.

The relative rating done in the laboratory has to be correlated with practice. This requires determination of wear rate in the sub-assembly test. The duration of such test depends on the accuracy with which wear can be assessed. As discussed earlier some of the newer approaches will be helpful in this regard. Also any comparison on the basis of steady state wear requires relatively longer durations. The sub-assembly tests can be conducted under several conditions. In the first instance the obvious choice is to run the tests under the most severe conditions expected in operation. The interest will be on the zone of maximum wear. If wear data for different formulations is available then comparison of the data is possible with the laboratory tests. Such a comparison will provide adequate confidence for laboratory evaluation. Identical wear rates are not to be expected as in the sub-assembly tests there can be mixed lubrication. For comparison all tests in sub-assembly have to be done with same conditions so that mixed lubrication effects are the same for all the additives. Though it is desirable to do sub-assembly tests under different operating conditions comparisons on the basis of the most severe conditions may be adequate.

The above approaches are no doubt time consuming. But as has been stated earlier the extent of wear mapping to be done has to be decided by the investigator and some simplifications are possible. It needs to be emphasised that wear mapping is not for the purpose of comparison alone. The maps amount to engineering modelling of boundary lubrication. They provide realistic information on all aspects of boundary lubrication that include influence of operating parameters, possible wear transitions, and running-in behaviour. When transitions occur in wear behaviour, empirical equations may have to be obtained for different zones.

The long-range research that can be conducted has been discussed in chapter 5 and is not considered here.

The comparisons are between fresh formulations, which may not undergo major deterioration in the laboratory tests. On the other hand the long duration test in the sub-assembly may involve lubricant deterioration, which can affect wear rates. To test this aspect some tests should be conducted on the oil samples periodically drawn from the sub-assembly test. This will clarify any adverse effects due to deterioration. Environment also should be kept in mind. For example the deterioration in an engine test will be in the presence of combustion products unlike the normal atmosphere in the laboratory test. The influence of acidic components generated in the combustion process on additives, as well as their direct interaction with materials is of importance. In diesel engines soot also plays a role in engine wear. Some examples of research in these areas may be cited [27, 28,29]. Correlations in such cases will depend on the extent of environmental influence and have to be carefully judged.

The progress in this direction eventually depends on the acceptance and use of the proposed methodology. As discussed in earlier chapters no clear demarcation between additives is possible with short duration tests. These tests have a history of more than five decades. The author was not successful in ordering this large information as it was not possible to define steady state wear rate satisfactorily due to reasons already discussed in chapter 7. Also different criteria were used to obtain wear rate and many comparisons between additives were at the laboratory level only. Information on correlation with practice is very limited. Even though mapping is initially expensive, in the longer run the advantages far outweigh the initial costs.

A question that arises is the extent to which wear map approach can help in evaluating novel formulations. The general feeling in the industry with testing in other areas like oxidation stability and detergency is that correlations are meaningful provided they have similar chemistry. This aspect is vague to define and may be considered to mean additives of similar class with variations in structure. Thus all ZDTP additives with different alkyl/aryl groups may be considered to have similar chemistry. The author considers that comparison on the basis of wear maps may be capable of relaxing this restriction and laboratory screening will be more meaningful even when all together different chemistry is involved.

Another relevant question concerns the user industry. The question is how to choose between different formulations. Large-scale users have facilities to do the relevant testing and also in some cases define their own requirements. They also influence the development of standards. With small scale users the choices are difficult. While independent laboratories can provide answers these may be expensive to come by. Such users may like to consider the idea of wear mapping. This should provide a better method to choose in comparison to short duration tests normally done.

The approaches suggested are logical but need to be tested. Verification from the available literature is not possible as comparisons based on wear map approach are not available. The author considers the suggested direction should be pursued by those involved in boundary lubrication research and practice. It is hoped the methodology will be used, and further developed in future.

References

1. H. Heizler, Vehicle and Engine Technology, Arnold, London, 1999.
2. C. M. Taylor (ed.), Engine Tribology, Tribology series, 26, 1993.
3. B. L. Ruddy, D. Dowson, and P. N. Economou, A review of studies of piston ring lubrication, in Tribology of Reciprocating Engines, paper V (i), Proc. 9th Leeds-Lyon Symposium on Tribology, 1982.
4. S. D. Gulwadi, Analysis of tribological performance of a piston ring pack, Trib. Trans., STLE, 43 (2000) 151.
5. C. M. Taylor, Fluid film lubrication in automobile valve trains, Proc. Instn. Mech. Engrs., J. Eng. Trib. Part J, 208 (1994) 221.
6. J. F. Booker, Dynamically loaded journal bearings: mobility method of solution, J. Basic Eng., ASME, D 187 (1965) 537.
7. C. M. Taylor, Automobile engine tribology-design considerations for efficiency and durability, Wear 221 (1), (1998) 1.
8. V. R. K. Sastry, S. D. Phatak, and A. Sethuramiah, Experimental study of the influence of liquid lubricants on burnished MoS_2 films, Wear, 86 (1983) 213.
9. M. Priest and C. M. Taylor, Automobile engine tribology-approaching the surface, Wear, 241 (2000) 193.
10. R. C. Coy, Practical applications of lubrication models in engines, New Directions in Tribology, MEP (1997) 197.
11. L. L. Ting, Lubricated piston ring and cylinder bore wear, in M. B. Peterson and W. O. Winer (eds.), Wear Control Handbook, ASME, 1980.

12. J. C. Bell and T. Colgan, Pivoted-follower valve train wear: Lub. Eng., STLE, 47 (2), (1991) 114.
13. A. Flodin and S. Andersson, Simulation of mild wear in helical gears, Wear, 241 (2000) 123.
14. M. C. Shaw, Metal Cutting Principles, Clarendon Press, Oxford, 1984.
15. P. Pawlus, A study on the functional properties of honed cylinder surfaces during running-in, Wear, 176 (1994) 247.
16. Y. R. Jeng, Impact of plateaued surfaces on tribological performance, Trib. Trans., STLE, 39 (1996) 354.
17. J. Michalski and P. Pawlus, Description of honed surface topography, Int. J. Mach. Tools, 34 (2), (1994) 199.
18. B. J. Taylor and T. S. Eyre, A review of piston ring and cylinder liner materials, Trib. Int., April (1979) 79.
19. S. M. Hill, S. E. Hartfield-Wunch, and S. C. Tung, Bench wear testing of common gasoline engine cylinder bore surface/piston ring combinations, Trib. Trans., STLE, 39 (1996) 929.
20. P. J. Blau, Friction and Wear Transitions of Materials, Noyes Publications, New Jersey, 1989, 337-351.
21. A. H. Naeim, M. Zheng, H. Shengqiang, B. Walter, and G. John, In situ wear measuring technique in engine cylinders, Trib. Trans., STLE, 41 (1998) 579.
22. R. Kumar, S. Kumar, B. Prakash, and A. Sethuramiah, Assessment of engine liner wear from bearing area curves, Wear, 239 (2000) 282.
23. R. Gahlin and S. Jacobson, A novel method to map and quantify wear on a micro-scale, Wear 222 (1998) 93.
24. K. Sasajima, K. Naoi, and T. Tsukuda, A software-based relocation tehnique for surface asperity profiles and its application to calculate volume changes in running-in wear, Wear, 240 (2000) 152.
25. E. Decenciere and D. Jeulin, Morphological decomposition of the surface topography of an internal combustion engine cylinder to characterize wear, Wear, 249 (2001) 482.
26. P. J. Guichelar, M. W. Williams, C. W. Omo, and D. Wilde, Experimental verification of thermoelastc mechanical seal face deflection by local wear measurements, Lub. Eng., 56 (7), (2000) 26.
27. Y. Yahagi, Y. Nagasawa, S. Hotta, and Y. Mizutani, Corrosive wear of cast iron under reciprocating lubrication, SAE 861599 (1986).
28. C. Kim, C. Passut, and D. M. Zang, Relationships among oil composition combustion generated soot, and diesel engine valve train wear, SAE 922199 (1992).
29. J. J. Truhan, C. B. Covington, and L. M. Wood, The classification of lubricating oil contaminants and their effect on wear in diesel engines as measured by surface layer activation, SAE 952558 (1995).

Nomenclature

b	depth of cut
k_b	wear factor, dimensional
l	sliding distance
m	exponent
n	exponent
p	exponent
t	feed rate
T_t	tool life
V	wear volume
v	cutting speed
W	load, N

Index